Data for Process Design and Engineering Practice

Donald R. Woods
McMaster University

PTR Prentice Hall
Englewood Cliffs, New Jersey 07632

ISBN 0–13–318149–9

Cover design: *Oysterpond Press*
Acquisitions editor: *Betty Sun*
Manufacturing manager: *Alexis A. Heydt*

Prentice-Hall, Inc.
A Paramount Communications Company
Englewood Cliffs, New Jersey 07632

The publisher offers discounts on this book when ordered in bulk quantities.
For more information, contact:

Corporate Sales Department
PTR Prentice Hall
113 Sylvan Avenue
Englewood Cliffs, NJ 07632

Phone: 201–592–2863
Fax: 201–592–2249

Printed in the United States of America
10 9 8 7 6 5 4 3 2 1

ISBN 0–13–318149–9

Prentice-Hall International (UK) Limited, *London*
Prentice-Hall of Australia Pty. Limited, *Sydney*
Prentice-Hall Canada Inc., *Toronto*
Prentice-Hall Hispanoamericana, S.A., *Mexico*
Prentice-Hall of India Private Limited, *New Delhi*
Prentice-Hall of Japan, Inc., *Tokyo*
Simon & Schuster Asia Pte. Ltd., *Singapore*
Editora Prentice-Hall do Brasil, Ltda., *Rio de Janeiro*

Contents

Preface

Process Design and Engineering Practice is not possible without data. The general principles for selecting process equipment cannot be applied unless we have the appropriate data. Some key data are the "units of measurement"; others pertain to conversion of concentrations. The properties of materials are also needed.

This is a companion to "Process Design and Engineering Practice", **PDEP**. Since Units of measurement are the bane of our existence, all of the units used in **PDEP** are in SI units. Hence, we need to be able to convert to and from SI to and from any set of units. Detail conversion factors and experience factors are given for most units in Part A. Another frustration I commonly encounter is conversion from the units for composition. "If the composition is 3 mol/L, what is the mole fraction in solution?". Part B addresses this issue.

To select options and size options we need values for the pertinent properties of gases, liquids and solids. The sloppiest and quickest approach is to use the single value approximations, given in Part A, under each type of measurement. However, most of us have memorized those values. To improve our estimates (and apply the principle of successive approximation), we need rapid access to approximate values of the key properties of gases, liquids and solids. Parts C and D provide such information. Too often we have to spend exorbitant amounts of time searching for the property data needed to quantify our ideas. I have tried to organize the data *in the format* that helps us select and size options. I have resisted the temptation to provide detailed correlations because, 1) they are available elsewhere (if I can find them) but more importantly, 2) I want to display the property information in a format such that, by scanning a table, you can estimate a reasonable value when the information is missing. This book is meant to be used. It is meant to be used and used effectively. If more data are needed, I have tried to identify resources that I would consult.

These data are published separately so that they can complement the other books in this series on Process Design and Engineering Practice. To facilitate the integration of this book with the companion books, the term **PDEP** is used in this book to refer to others; this book is referred to as **Data** in those companion books.

How will you benefit from this text and how might you use it? This provides an extremely convenient collection of the data needed for process design. I know of no other source that attempts to provide as much data, in consistent sets of units and organized in a format for use. How do you use it? As a companion for any engineering activity. It provides key information freshman need in early design projects; the section on unit conversions and practical experience numbers will prove to be valuable. "Mass and energy balances is OK, but converting the units to solve the problem is a pain". Sound familiar? Parts A and B are meant to eliminate some of the pain. Problems in fluid mechanics, heat transfer and selecting separation options require the information in Parts C and D.

This book is meant to be used. Blank spaces in the Tables are meant to be filled in. Add your own values as they are discovered. Just as this book is an outgrowth of the needs I had as a practicing engineer, so may this be a basis for you in your profession.

Donald R. Woods
Waterdown, Ontario

Part A: Units and Conversion of Units.

Here are conversion factors for about 70 commonly-used units of measurement. All are expressed in terms of the standard SI units, with conversion factors from the myriad of systems of units used. (For example, a "herring tub" is an official unit of measurement in Canada. What's that in SI? Actually, 72.73795 dm^3. However, since I have not encountered a need to use this value for PDEP over the past 30 years, I have not included it here.) Included are all of the unit conversions I had to work my way through. I trust I included the ones you need. Also included for most units are "experience" values. The units are organized by "dimension" starting with length L and progressing through to force, then energy and then electrical and magnetic units. The index can help you locate the values rapidly.

SI Prefixes:

10^{18}	exa E	10^{-1}	deci d
10^{15}	peta P	10^{-2}	centi c
10^{12}	tera T	10^{-3}	milli m
10^{9}	giga G	10^{-6}	micro μ
10^{6}	mega M	10^{-9}	nano n
10^{3}	kilo k	10^{-12}	pico p
10^{2}	hecto h	10^{-15}	femto f
10	deka da	10^{-18}	atto a
		10^{-21}	beta β

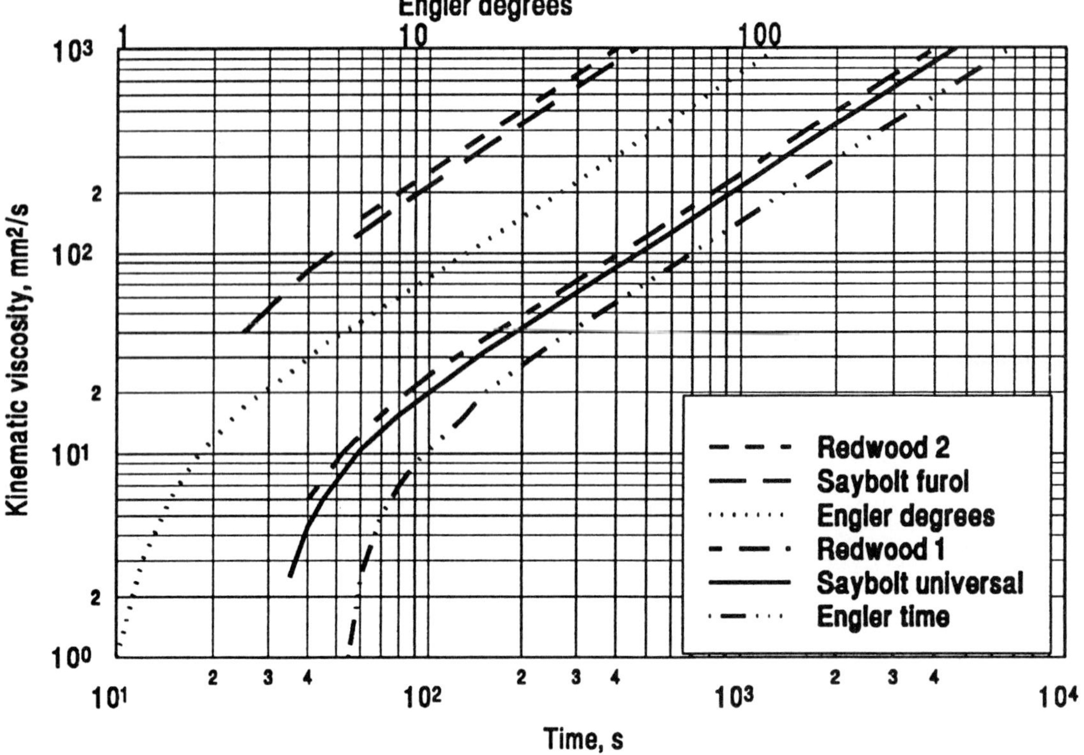

Figure A–1 Conversions Among Systems of Units for Kinematic Viscosity of Liquids

CONVERSIONS for DISTANCE, LENGTH	dimensions L		
metre, m	inches	x 0.0254	= m
height of adult: 1.6 to 1.8 m	ft	x 0.3048	= m
height of giraffe: 10 m	yard	x 0.914	= m
	miles	x 1.609	= km
	µinch	x 0.0254	= µm
	mil	x 0.00254	= cm
	cm	x 0.01	= m
	angstrom	x 10^{-10}	= m
	micron	x 10^{-6}	= m

CONVERSIONS for PERMEABILITY	dimensions L^2		
square micrometre, μm^2	D'Arcy	x 0.9869	= μm^2
	ft^3lb_m/h^2lb_f	x 222.62	= μm^2
28 μm filter: 10 μm^2	ft^3lb_m/s^2lb_f	x 2.885E9	= μm^2
clay: 0.01 μm^2			

CONVERSIONS for AREA, LENGTH	dimensions L^2		
square metre, m^2	$inches^2$	x 6.45E-4	= m^2
	ft^2	x 0.0929	= m^2
card table: 0.6 m^2	$yard^2$	x 0.836	= m^2
	$miles^2$	x 2.5899	= km^2
	acres	x 4047	= m^2
	dm^2	x 10^{-2}	= m^2
	cm^2	x 10^{-4}	= m^2
	mm^2	x 10^{-6}	= m^2
	$angstrom^2$	x 10^{-2}	= nm^2
	$micron^2$	x 10^{-12}	= m^2
	km^2	x 10^6	= m^2
	hectare (ha)	x 10^4	= m^2

CONVERSIONS for VOLUME	dimensions: L^3		
cubic metre, m^3	$inches^3$	x 16.39E-6	= m^3
	ft^3	x 28.2E-3	= m^3
box car: 100 m^3	$yard^3$	x 0.7646	= m^3
refrigerator: 1 m^3	Imp. gallon	x 4.55	= L
large pail: 10 L	US gallon	x 3.785	= L
brick: 1 L	Imp. quart	x 1.1365	= L
golf ball: 40 cm^3	kL	x 1	= m^3
1 m^3 ~ 1 tonne water	barrel (oil)	x 0.15899	= m^3
	fluid oz	x 28.413E-3	= L
	pint	x 0.568	= L
	std ft^3 (STP)	x 28.2E-3	= Nm^3 (STP)
	std ft^3 (STD)	x 26.7E-3	= Nm^3 (STP)
standard conditions:	std ft^3 (STP)	x 29.7E-3	= m^3 (STD)
STP gas = 0°C, 101.325 kPa dry; metric uses the prefix	dm^3 = L	x 10^{-3}	= m^3
"N" to designate this condition.	hectolitre	x 0.10	= m^3
Other "standard" conditions used include:	cm^3	x 10^{-6}	= m^3
STD gas = 15.6°C, 101.325 kPa dry	cm^3	x 1	= mL
NTP(API) = 15°C, 101.325 kPa	mm^3	x 10^{-9}	= m^3
	nm^3	x 10^{-27}	= m^3
	$micron^3$ = μm^3	x 10^{-18}	= m^3
	$(km)^3$	x 10^9	= m^3
	$(hectom)^3$	x 10^6	= m^3
	$(decam)^3$	x 10^3	= m^3

CONVERSIONS for VOLUME RATIO	dimensions: dimensionless		
	scfm/100 USgpm	x 7.45	= dm^3/100L
	USgal/1000 acf	x 0.134	= L/m^3
	USgal/bbl.	x 0.0238	= m^3/m^3

CONVERSIONS for AMOUNT OF SUBSTANCE/VOLUME	dimensions: 1/L		
kmol/m^3	mol/L	x 1	= kmol/ m^3
	lb mol/ft^3	x 16.02	= kmol/ m^3
pure gaseous CO_2 at STP: 0.0446 kmol/m^3			
1 kmol = 22.4 m^3 at STP			

CONVERSIONS for VOLUMETRIC FLOW	dimensions: L^3/T		
cubic metres per second, m^3/s	ft^3/s	x 28.317	= dm^3/s
cubic decimetre per second, dm^3/s or L/s	$ft^3/$ min	x 0.4719	= dm^3/s
	scfm	x 1.6699	= m^3/h (STP)
fast running tap into a sink: 0.1 L/s	scfm	x 1.5	= Nm^3/h
liquid pumped through a 5 cm diam pipe: 2.5 L/s	scfm	x 0.4719	= dm^3/s(STP)
	scfh	x 0.028	= m^3/h (STP)
gas flowing through 10 cm diam pipe: 150 dm^3/s	scfh	x 7.756	= cm^3/s
	10^6 scfd	x 1.17E3	= m^3/h (STP)
	US gpm	x 0.0631	= L/s
	Imp. gpm	x 0.07577	= L/s
	Imp gph	x 1.26E-6	= m^3/s
	10^6 Imp. gpd	x 0.0526	= m^3/s
	10^6 US gpd	x 0.438	= m^3/s
	10^3 bbl/d	x 1.84	= L/s
	10^3 bbl/d	x 6.62	= m^3/h
	bbl/d	x 0.159	= m^3/d
	mm^3/s	x 10^{-9}	= m^3/s
	cm^3/s	x 10^{-6}	= m^3/s
	dm^3/s = L/s	x 10^{-3}	= m^3/s
	L/s	x 3.6	= m^3/h
	L/min	x 1.667E-5	= m^3/s
	m^3/h	x 2.778E-4	= m^3/s

CONVERSIONS for ACCELERATION	dimensions: L/T^2		
m/s^2	cm/s^2	x 10^{-2}	= m/s^2
	ft/s^2	x 0.3048	= m/s^2
acceleration of gravity ~ 9.8 m/s^2			
std. 9.80665 m/s^2			

CONVERSIONS for ANGULAR ACCELERATION	dimensions: $1/T^2$
radians per second, r/s²	

CONVERSIONS FOR VELOCITY	dimensions: L/T		
Volumetric flowrate per unit area, volume flux, mass transfer coefficient for a concentration driving force, mass flux per unit concentration driving force			
metre per second, m/s	ft/s	x 0.3048	= m/s
	ft/min	x 0.00508	= m/s
liquid pumped *1 m/s*	ft/min	x 5.08	= L/m²s
	cfm/ft²	x 0.00508	= m/s
highway driving *25 m/s*	ft³/ft² min	x 5.08	= dm³/m²s
	US gal/ft² day	x 0.04074	= m³/m² day
Mass transfer coefficient	US gal/ft² day	x 4.715E-4	= L/m² s
	US gal/ft² h	x 0.04074	= m³/m² h
for gases *85 μm/s*	US gal/ft² h	x 0.011316	= L/m² s
for liquids *0.85 μm/s*	US gal/ft² min	x 2.45	= m³/m² h
	US gal/ft² min	x 0.6789	= L/m² s
	miles/h	x 1.6093	= km/h
	miles/h	x 0.447	= m/s
	ft³/acre s	x 0.06998	= m³/(ha.s)
	ft³/acre s	x 6.998E-6	= m/s
	mpy (mil per year)	x 0.025	= mm/a
	cm/s	x 10^{-2}	= m/s
	cm/min	x 0.166	= L/m² s
	cm³/cm² h	x 0.0028	= L/m² s
	dm/s	x 10^{-1}	= m/s
	m/min	x 1.6667E-2	= m/s
	m/h	x 2.7778E-4	= m/s
	mm/s	x 1	= L/m² s
	L/m² s	x 3.6	= m³/m² h

CONVERSIONS FOR VOLUMETRIC FLOW PER UNIT LENGTH dimensions: L^2T			
Kinematic viscosity, thermal or molecular diffusivity			
Square metre per second, m^2/s	US gal/(day ft)	x 0.01242	= m^3/day.m
	m^3/(day.m)	x 1.157E-5	= m^2/s
Diffusivities:	centistokes		= mm^2/s
for gases 0.1 to 1 cm^2/s	(1 stoke = cm^2/s)	x 10^{-4}	= m^2/s
	ft^2/s	x 0.0929	= m^2/s
for liquids 1000 $\mu m^2/s$	m^2/h	x 2.7778E-4	= m^2/s
	ft^2/h	x 2.5806E-5	= m^2/s
for solids 0.1 to 10^4 nm^2/s	in^2/s	x 6.451	= m^2/s
	dm^2/s	x 10^{-2}	= m^2s
Kinematic viscosity:	mm^2/s	x 10^{-6}	= m^2s
	cm^2/s	x 10^{-4}	= m^2/s
for water 10^6 $\mu m^2/s$	cm^2/s	x 10^8	= $\mu m^2/s$
	$\mu m^2/s$	x 10^{-12}	= m^2/s
	nm^2/s	x 10^{-18}	= m^2s
	SSU	x 2.165E-3	= cm^2/s

CONVERSIONS for MASS	dimensions: M		
kilogram, kg	lb_m	x 0.4536	= kg
	grain	x 6.48E-2	= g
metric ton = Mg	ton (2000 lb_m)	x 0.9072	= Mg
	long ton (2240 lb_m)	x 1.016	= Mg
your mass 50 to 90 kg	metric tonne	x 1.000	= Mg
	g	x 10^{-3}	= kg
	mg	x 10^{-6}	= kg

CONVERSIONS for MOLAR MASS	dimensions: M
kg/kmol **depends on the substance** **g/mol** *for water: molar mass is 18 kg/kmol* *for air: molar mass is 29 kg/kmol*	

CONVERSIONS for AMOUNT OF SUBSTANCE	dimensions: dimensionless		
kmol: mass of a substance divided by the molar mass	lb-mole	x 0.4536	= kmol
	mol	x 10^{-3}	= kmol
1 mol of He = 1 gram-atom of He			
1 mol of Na_2CO_3 = 1 gram-molecule of Na_2CO_3	DO NOT USE:		
1 mol of CLO^-_4 = 1 gram-ion of CLO^-_4	g-mole or kg-mole;		
1 mol of Cl_2 = mass of 70.914 g	**instead use:**		
1 mol of Cl^- = mass of 35.457 g	mol or kmol, respectively.		
1 mol of $^1/_2\ Ca^{2+}$ = mass of 20.04 g			
1 mol of e^- = mass of 548.6 μg			

CONVERSIONS for MASS RATIO	dimensions: dimensionless		
	$grains/lb_m$ dry air	x 0.1429	= g/kg

CONVERSIONS for MASS PER UNIT VOLUME	dimensions: M/L^3		
Density and mass concentration			
kilogram per cubic metre, kg/m^3	$slug/ft^3$	x 515.38	= kg/m^3
	lb_m/ft^3	x 16.02	= kg/m^3
density of liquid water = 1 Mg/m^3	$lb_m/1000\ ft^3$	x 0.01602	= kg/m^3
of air (STP) = 1.2 kg/m^3	lb_m/UK gal	x 99.779	= kg/m^3
of steel = 7.8 Mg/m^3	lb_m/US gal	x 119.8	= kg/m^3
of mercury = 13.6 Mg/m^3	lb_m/bbl	x 2.853	= kg/m^3
	grains per US gal	x 17.12	= g/m^3
	$grains/ft^3$	x 2.288	= g/m^3
	g/cm^3	x 10^3	= kg/m^3
	g/L	x 1	= kg/m^3
	mg/L	x 10^{-3}	= kg/m^3

CONVERSIONS for MOLAR CONCENTRATION	dimensions: $1/L^3$		
kilomol per cubic metre, $kmol/m^3$	mol/L	x 1	= $kmol/m^3$
DO NOT USE: molarity or "M" for molar solution. Another obsolete term is Molality: dimensions 1/M. This would have units of measurment of mol/kg			

CONVERSIONS for MASS/UNIT AREA	dimensions: M/L^2		
kilogram per square metre, kg/m^2	lb_m/ft^2	x 4.882	= kg/m^2
	lb_m/in^2	x 703.069	= kg/m^2
	short tons/acre	x 0.224	= kg/m^2
	short tons/sq mile	x 350.265	= mg/m^2

CONVERSIONS for SURFACE AREA PER UNIT MASS	dimensions: L^2/M		
square metre per kilogram: m^2/kg	cm^2/g	x 10^{-3}	= m^2/kg
	m^2/g	x 10^3	= m^2/kg
	ft^2/lb_m	x 0.2048	= m^2/kg

CONVERSIONS for MASS FLOWRATE	dimensions: M/T		
kilogram/second: kg/s	lb_m/s	x 0.4536	= kg/s
	lb_m/min	x 7.56	= g/s
	lb_m/h	x 126	= mg/s
	lb_m/h	x 0.4536	= kg/h
	MM lb_m/a	x 0.4536	= Gg/a
	short ton/h	x 0.28224	= kg/s
	short ton/day	x 0.30	= Mg/a(330 day)
		x 0.33	= Mg/a(365 day)
	kg/h	x 2.7778E-4	= kg/s

CONVERSIONS for MASS FLOWRATE PER UNIT LENGTH	dimensions: M/LT		
Viscosity			
Pascal. second: kg/(s•m) = Pa•s = N•s/m^2	lb_m/(s.ft)	x 1.4882	= Pa•s
	lb_m/(h.ft)	x 0.41338	= mPa•s
viscosity of liquid water: 1 mPa•s	lb_f•s/ft^2	x 47.88	= Pa•s
	poundal•s/ft^2	x 1.488	= Pa•s
Poiseuille = Pa•s	g/(cm•s)	x 0.1	= Pa•s
	Poise	x 0.1	= Pa•s
	Centipoise	x 1	= mPa•s
	PI	x 10^3	= mPa•s
	kg/(m•h)	x 0.2777	= mPa•s

CONVERSIONS for MASS FLOWRATE PER UNIT AREA	dimensions: M/L^2T		
Mass flux and mass transfer coefficient per unit dimensionless driving force.			
kilogram per second per square metre: kg/(s m^2)	lb_m/(s.ft^2)	x 4.882	= kg/(s.m^2)
	lb_m/(min.ft^2)	x 0.0814	= kg/(s.m^2)
	lb_m/(h.ft^2)	x 1.356	= g/(s.m^2)
	lb_m/(day.ft^2)	x 0.0565	= g/(s.m^2)
	ton/(h.ft^2)	x 2.712	= kg/(s.m^2)
	ton/(day.ft^2)	x 0.113	= kg/(s.m^2)
	g/(s.cm^2)	x 10	= kg/(s.m^2)
	kg/(h.m^2)	x 0.27778	= g/(s.m^2)

CONVERSIONS for MASS FLOWRATE PER UNIT VOLUME	dimensions: M/L^3T		
Reaction rate			
kilogram per second per cubic metre: kg/(s m^3)	lb_m/(s.ft^3)	x 16.018	= kg/(s.m^3)
	lb_m/(h.ft^3)	x 4.45	= g/(s.m^3)
liquid phase reaction rate: 10 to 200 g/s.m^3	lb_m/(day. 1000 ft^3)	x 0.01602	= kg/(s.m^3)
	g/(s.cm^3)	x 1	= Mg/(s.m^3)
	kg/(h.m^3)	x 0.27778	= g/(s.m^3)

CONVERSIONS for MASS FLUX PER UNIT PRESSURE DRIVING FORCE dimensions: T/L			
Mass transfer coefficient per unit pressure driving force.			
kilogram per second per square metre per kilopascal driving force: kg/(s•m^2 •kPa)	lb_m/(h.ft^2•atmos.)	x 13.384	= mg/(s.m^2•kPa)
	g/(s.cm^2•atmos.)	x 0.09869	= kg/(s.m^2•kPa)
	g/(s.cm^2•atmos.)	x 98.89	= kg/(s.m ^{2} •MPa)
gas film controlled: 10 mg/s•m^2•kPa	kg/(h.m^2•atmos.)	x 2.741	= mg/(s.m^2•kPa)

CONVERSIONS for VOLUME FLUX PER UNIT PRESSURE DRIVING FORCE dimensions: LT^2/M			
cubic metres per second per square metre per kiloPascal driving force: m^3/(s.m^2.kPa)	US gal/ft^2.day.psi)	x 0.0706	= mL/m ^{2} s.MPa

CONVERSIONS for FORCE	dimensions: ML/T^2		
Newton, N = kg.m/s^2	lb_f	x 4.448	= N
force acting on an apple in the earth's gravitational field is about 1 N;	poundal	x 0.138	= N
	(2000 lb_f) ton_f	x 9.96	= k N
	kip	x 4.448	= kN
g = acceleration due to "standard" gravity = 9.8067 m/s^2	kg_f	x 9.8067	= N
	dyne	x 10^{-7}	= N
In a gravitational field of "g", the force an a mass M is Mg Newtons.			

CONVERSIONS for MOMENTUM	dimensions: ML/T		
kilogram metre per second: kg.m/s	lb_m.ft/s	x 0.138225	= kg.m/s
	lb_m.ft/h	x 3.8396E-5	= kg.m/s
	g.cm/s	x 10^{-5}	= kg.m/s

CONVERSIONS for ANGULAR MOMENTUM	dimensions: ML^2/T		
kilogram square metre per second: kg.m^2/s	lb_m.ft^2/s	x 0.042145	= kg.m^2/s
	lb_m.ft^2/h	x 1.1706E-5	= kg.m^2/s
	g.cm^2/s	x 10^{-7}	= kg.m^2/s

CONVERSIONS for MOMENT OF INERTIA	dimensions: ML^2		
kilogram square metre: kg.m^2	lb_m.ft^2	x 0.042145	= kg.m^2
	lb_m.in^2	x 2.926E-4	= kg.m^2
	slug.ft^2	x 1.355	= kg.m^2
	kg.cm^2	x 10^{-4}	= kg.m^2

CONVERSIONS for FORCE PER UNIT AREA — dimension: M/LT^2

Pressure, stress, momentum flowrate per unit area

kilopascal, kPa

Pressure is always absolute *(not gauge pressure)*

≡ 1000 N/m^2
≡1/100 bar
≃1/100 atmos.
≃blow into a manometer and display 10 cm water vertically.

$Pa = kg.s^{-2}.m^{-1}$

Atmospheric pressure: 101.325 kPa, 760 mm mercury, 29.921 inches mercury

Stress:

strength of concrete	*20 MPa*
design stress of concrete	*9 MPa*
yield strength of steel	*350 MPa*
design stress of steel	*165 MPa*

atmos	x 101.33	= kPa
bar	x 100	= kPa
psi	x 6.895	= kPa
inches water (at 3.9°C)	x 0.249	= kPa
feet water	x 2.989	= kPa
cm water	x 0.098	= kPa
inches mercury (at 0°C)	x 3.386	= kPa
mm mercury	x 0.133	= kPa
torr	x 0.133	= kPa
micron	x 0.133	= Pa
N/m^2	x 10^{-3}	= kPa
N/mm^2	x 1	= MPa
kg_f/cm^2	x 98.066	= kPa
lb_f/ft^2	x 47.88	= kPa
$dyne/cm^2$	x 0.1	= Pa
$g/cm.s^2$	x 0.1	= Pa
ksi	x 6.895	= MPa
$tons_f/in^2$	x 15.4	= MPa

CONVERSIONS for PRESSURE/UNIT LENGTH — dimensions: M/L^2T^2

kPa/metre:

pressure drop for gas flowing through packed bed: 20 kPa/m

in water/ft	x 0.82	= kPa/m

CONVERSIONS for ENERGY	dimension: ML^2/T^2		
Torque, moment of force.			
Joule, J ≡ N.m ≡ W.s	cal	x 4.187	= J
	BTU	x 1.055	= kJ
Joule is the amount of energy to raise the temperature of a cube of water 6 mm x 6 mm x 6 mm by 1 °C.	hp.hr	x 2.684	= MJ
	kW.h	x 3.6	= MJ
	MM BTU	x 1.055	= GJ
	10^6 kcal	x 4.187	= GJ
	kg_f.m	x 9.8066	= J
	ft.lb_f	x 1.3558	= J
bbl crude oil equiv. x 5.9 = GJ	in.lb_f	x 0.133	= J
Mg crude oil equiv. x 44 = GJ	in.oz_f	x 7.062	= mJ
m^3 crude oil equiv. x 37 = GJ	dyne.cm	x 10^{-7}	= J
Mg coal equiv. x 29 = GJ	erg	x 10^{-7}	= J
m^3 natural gas equiv. x 37 = MJ	erg	x 0.1	= μJ
ft^3 natural gas equiv. x 1.05 = MJ	electron volt	x 0.1602	= aJ
Mg deuterium (fusion equiv.) x 0.25 = EJ	therm	x 105.5	= MJ
Mg uranium 235(fission eq.) x 83 = PJ	CHU or PCU	x 1.899	= kJ
ton (nuclear equiv. of TNT) x 4.2 = GJ	Quad	x 1.055	= EJ
ton-day of refrigeration x 0.5275 = MJ	litre-atmos.	x 0.1011	= kJ
	ft^3-atmos.	x 2.869	= kJ
	poncelet-h	x 3.353	= MJ

CONVERSIONS for ENERGY/UNIT AMOUNT OF SUBSTANCE	dimensions: ML^2/T^2		
kilojoule per kilomole: kJ/kmol:	cal/mol	x 4.1868	= kJ/kmol
	BTU/lb-mole	x 2.326	= kJ/kmol
latent heat of evaporation for water: 45 MJ/kmol			
latent heat of fusion for water: 6 MJ/kmol			
typical heat of reaction: 200 MJ/kmol			

CONVERSIONS for FORCE/UNIT LENGTH or ENERGY PER UNIT AREA dimensions: M/T^2			
Surface tension, surface energy per unit area.			
Newton per metre: N/m or	dyne/cm	x 1	= mN/m
joule per square metre: J/m^2	lb_f/ft	x 14.59	= N/m
surface tension for water: 72 mN/m	erg/cm^2	x 10^{-3}	= J/m^2
for oils: about 20 mN/m	erg/cm^2	x 1	= mJ/m^2
	mN/m	x 1	= mJ/m^2

CONVERSIONS for ENERGY/UNIT VOLUME	dimensions: M/T^2L		
Hildebrand solubility parameters			
Joule per cubic metre: kJ/m^3	cal/cm^3	x 4.1868	= MJ/m^3
heating values for:	$kcal/m^3$	x 4.1868	= kJ/m^3
low quality synthetic gas: 15 MJ/m^3	BTU/ft^3	x 37.26	= kJ/m^3
natural gas: 35 MJ/m^3	$therm/ft^3$	x 3.726	= GJ/m^3
No. 2 fuel oil: 42 MJ/L	1000 BTU/bbl.	x 6.636	= MJ/m^3
diesel oil: 42 MJ/L	kWh/bbl.	x 22.643	= MJ/m^3
bunker C fuel oil: 42 MJ/L	hp/1000 cfm	x 1.58	= $kW.s/m^3$
Hildebrand solubility parameters:	1000 BTU/UK gal	x 0.2318	= MJ/L
dispersion contribution: 10 $(J/cm^3)^{1/2}$	$(cal/cm^3)^{1/2}$	x 2.046	= $(J/cm^3)^{1/2}$

CONVERSIONS for ENERGY/UNIT MASS	dimension: L^2/T^2		
Latent heat, specific enthalpy; reciprocal: mass of gas absorbed per energy input.			
kilojoule per kilogram, kJ/kg	cal/g	x 4.1868	= kJ/kg
	BTU/lb_m	x 2.326	= kJ/kg
Latent heat of water: *2000 kJ/kg*	kW.h/lb_m	x 7775.7	= MJ/Mg
of organics: *400 kJ/kg*	CHU or PCU/lb_m	x 4.1868	= kJ/kg
heat of fusion of water: *300 kJ/kg*	reciprocal:		
of organics: *125 kJ/kg*	lb_m/(hp.h)	x 0.169	= mg/J
	lb_m/(hp.h)	x 0.6084	= kg/(kW.h)

CONVERSIONS for ENERGY/DEGREE/UNIT AMOUNT OF SUBSTANCE dimensions: $ML^2/T^2\theta$			
Molar entropy, molar heat capacity			
kilojoule per kilomole per degree Kelvin: kJ/(kmol.K):	cal/mol.°C	x 4.1868	= kJ/(k mol.K)
	BTU/lb-mole.°F	x 4.1868	= kJ/(kmol.K)
molar heat capacity:			
liquid water: *75 kJ/(kmol.K)*			
air: *30 kJ/(kmol.K)*			

CONVERSIONS for ENERGY/UNIT DEGREE/UNIT MASS	dimensions: $L^2/T^2\theta$		
Specific heat capacity			
kilojoule per kilogram per degree Kelvin: kJ/(kg.K)	cal/g.°C	x 4.1868	= kJ/(kg.K)
	BTU/lb_m.°F	x 4.1868	= kJ/(kg.K)
heat capacity for liquid water: *4.2 kJ/(kg.K)*			
for gas water: *2 kJ/(kg.K)*			
for air: *1 kJ/(kg.K)*			

CONVERSIONS for ENERGY/UNIT TIME	dimensions: ML^2/T^3		
Power, energy flowrate, energy "duty"			
kilowatt: kW ≡ kJ/s ≡ kg.m²/s³	cal/s	x 4.187	= W
	kcal/h	x 1.163	= W
Boiler horsepower: 34.5 lb_m of water evaporated per hour of dry, saturated steam at 100°C	tonne-cal/h	x 1.163	= kW
	BTU/s	x 1.055	= kW
	kBTU/h	x 0.2931	= kW
	hp	x 0.7457	= kW
	kg_f.m/s	x 9.8066	= W
	ft.lb_f/s	x 1.3558	= W
	erg/s	x 10^{-7}	= W
	MJ/h	x 0.277	= kW
	cheval	x 0.736	= kW
	ton refrigeration	x 3.5169	= kW

CONVERSIONS for ENERGY/UNIT TIME/UNIT AMOUNT OF SUBSTANCE dimensions: ML^2/T^3			
Energy flowrate/ amount of substance			
kilowatt per kilomole: kW/kmol	BTU/lb-mole.hr	x 0.646	= W/kmol

CONVERSIONS for ENERGY/UNIT TIME/UNIT MASS	dimensio ns: L^2/T^3		
Energy flowrate/unit mass			
kilowatt per kilogram: kW/kg	cal/(s.g)	x 4.1868	= kW/kg
	kcal/(h.kg)	x 1.163	= W/kg
	kBTU/(h.lb_m)	x 0.646	= kW/kg

CONVERSIONS for ENERGY FLOWRATE/UNIT AREA	dimensions: M/T^3		
Heat flux density, energy flux			
kilowatt per square metre: kW/m^2:	cal/(s.cm^2)	x 41.868	= kW/m^2
	kcal/(h.m^2)	x 1.163	= W/m^2
radiant heat transfer: 40 to 60 kW/m^2	kBTU/(h.ft^2)	x 3.1546	= kW/m^2
critical boiling heat flux:	MJ/(h.m^2)	x 0.277	= kW/m^2
for water 1000 kW/m^2	kJ/(h.m^2)	x 0.277	= W/m^2
for organics 90 kW/m^2			

CONVERSIONS for ENERGY FLOWRATE/UNIT VOLUME	dimension: M/LT^3		
Volumetric heat release rate, power per unit volume			
watt per cubic metre: W/m^3	cal/(s.cm^3)	x 4.1868	= MW/m^3
	kcal/(h.m^3)	x 1.163	= W/m^3
Turbine mixing in a tank of liquid 1 kW/m^3	kBTU/(h.ft^3)	x 10.35	= kW/m^3
	kcal/(h.ft^3)	x 41.07	= W/m^3
	hp/1000 US gal	x 0.1997	= kW/m^3
	hp/1000 ft^3	x 0.0264	= kW/m^3
	ft.lb_f/(s.ft^3)	x 0.048	= kW/m^3

CONVERSIONS for ENERGY FLUX PER DEGREE DRIVING FORCE	dimensions: $M/T^3\theta$		
Heat transfer coefficient			
watts per square metre per degree Kelvin: W/(m^2.K):	cal/(s.cm^2.K)	x 41.868	= kW/m^2.K
	kcal/(h.m^2.K)	x 1.163	= W/m^2.K
	kBTU/(h.ft^2.°F)	x 5.6784	= kW/m^2.K
condensing steam: 5000 W/m^2.K	CHU/(h.ft^2.K)	x 5.6784	= W/m^2.K
gas-gas transfer: 25 W/(m^2.K)	PCU/(h.ft^2.K)	x 5.6784	= W/m^2.K
	lb-cal/(h.ft^2.°C)	x 5.6784	= W/m^2.K
	cal/(h.cm^2.°C)	x 0.01163	= kW/m^2.K

CONVERSIONS for ENERGY FLUX PER DEGREE PER UNIT LENGTH	dimensions: $ML/T^3\theta$		
Thermal conductivity			
watts per square metre per degree Kelvin per metre: W/(m^2.K/m) ≡ W/(m.K)	cal/(s.cm.°C)	x 418.68	= W/(m.K)
	kcal/(h.m.°C)	x 1.163	= W/(m.K)
	BTU/(h.ft°F)	x 1.7308	= W/(m.K)
thermal conductivity of	BTU/(h.ft^2.°F/in)	x 0.1442	= W/(m.K)
liquid water: 0.6 W/(m.K)			
gas: 20 mW/(m.K)			

CONVERSIONS for ENERGY FLOWRATE/UNIT VOLUME PER DEGREE DRIVING FORCE dimensions: $M/LT^3\theta$

Volumetric heat transfer coefficient

watts per cubic metre per degree Kelvin: $W/(m^3.K)$

$CHU/(h.ft^3.°C)$	x 18.63	$= W/(m^3.K)$
$BTU/(h.ft^3.°F)$	x 18.63	$= W/(m^3.K)$

CONVERSIONS for CHARGE dimensions: IT

Coulomb, C

large value	*1 C*
medium value	*100µC*
small value	*50 pC*
electrostatic charge on a person:	*100 nC*
charge on an electron:	*0.16 aC*
charge transfer in/out digital CMOS transistor:	*10 pC*
charge on 1000 µF capacitor at 10 V	*10 mC*

statcoulomb	x 0.3336	= nC
abcoulomb	x 10	= C
faraday		
(based on C^{12})	x 96.487	= kC
(chemical)	x 96.4957	= kC
(physical)	x 96.5219	= kC
ampere-hr	x 3600	= C

CONVERSIONS for CHARGE PER UNIT AREA dimensions: IT/L^2

Surface charge density, electric derived field or displacement, dielectric polarization

Coulomb per square metre, C/m^2

clay particles in water	*$40mC/m^2$*
small charge	*$\mu C/m^2$*
dielectric capacitors:	*mC/m^2*
on control electrode in CMOS device:	*mC/m^2*

$statcoulomb/cm^2$	x 3.336	$= \mu C/m^2$
$abcoulomb/cm^2$	x 0.1	$= MC/m^2$
statvolt/cm	x 0.2653	$= \mu C/m^2$
abvolt/cm	x 7.96	$= kC/m^2$

CONVERSIONS for CHARGE PER UNIT VOLUME dimensions: IT/L^3

Volumetric charge density.

Coulomb per cubic metre, C/m^3

conduction electric charge	
in metal conductors:	*$16\ GC/m^3$*
in n-type semiconductors:	*$1.6\ kC/m^3$*

$statcoulomb/cm^3$	x 0.3336	$= mC/m^3$
$abcoulomb/cm^3$	x 10	$= MC/m^3$

CONVERSIONS for CHARGE FLOWRATE dimensions: I

Current

Ampere: A

start a 0.25 kW motor	*13 A*
electric kettle	*13 A*
refine alumina	*10 kA*
usual current in low power junction type semiconductor	*10 µA to 10 mA*
usual control electrode or OFF current in CMOS devices:	*pA to nA*
usual current in SCR power devices:	*1A to 1 kA*
dangerous lethal current inside human bodies:	*100 µA to 1 mA*

statamp	x 0.33356	= nA
abamp	x 10	= A
esu current	x 0.3336	= nA
emu current	x 10	= A
amp-turn	x 10	= A
gilbert	x 0.7958	= A

CONVERSIONS for CHARGE FLOWRATE PER UNIT LENGTH	dimensions: I/L		
Magnetism, magnetic derived strength, magnetic intensity			
Ampere per metre, A/m	esu	x 2.653	= nA/m
	gauss	x 79.577	= A/m
for iron *1 kA/m*	oersted	x 79.577	= A/m
	abampere-turn/cm	x 1	= kA/m
	amp-turn/cm	x 100	= A/m
	amp-turn/in	x 39.37	= A/m
	praoerstead	x 4 π	= A/m

CONVERSIONS for CHARGE FLOWRATE PER UNIT AREA	dimensions: I/L^2		
Current density, current per unit area, charge flux			
ampere per square metre, A/m^2	statamp/cm^2	x 3.336	= μA/m^2
	abamp/cm^2	x 100	= kA/m^2
electroplating or electrorefining: 10 to 2000 A/m^2	amp/ft^2	x 10.87	= A/m^2
semiconductor junction: 10 MA/m^2			
solar cell: 400 A/m^2			

CONVERSIONS for ELECTRIC POTENTIAL	dimensions: ML^2/IT^3		
Electromotive force			
Volt, V = J/A.s	statvolt	x 299.79	= V
Power transmission lines: 250 kV, 500kV	esu electric pot.	x 299.79	= V
household circuits:	abvolt	x 10	= nV
North American 110 V (rms)	emu electric pot.	x 0.01	= μV
European & Asian 240 V (rms)	International		
clay particles suspended in water: 80 mV	(1948) volt	x 1.000330	= V
magnetic phono cartridge			
signal output level: 5 mV			
muscle (EMG) cell potential: 1 mV			
electrostatic potential:			
on human body: 1 kV			
accelerating voltage			
in TV picture tube: 30 kV			

CONVERSIONS for ELECTRIC POTENTIAL PER UNIT LENGTH	dimensions: ML/IT^3		
Potential gradient			
Volt per metre, V/m = J/(m.A.s)	statvolt/cm	x 29.979	= kV/m
	abvolt/cm	x 1	= μV/m
breakdown occurs in air: 2 MV/m			
breakdown of semiconductors: 20 MV/m			
radiated field near a transformer: 1 mV/m			

CONVERSIONS for RESISTANCE		dimensions: ML^2/I^2T^3	
Ohm, Ω = V/A	statohm	x 898.76	= GΩ
	esu resistance	x 898.76	= GΩ
leakage resistance of	abohm	x 1	= nΩ
small telflon stand-off: 10^{14} Ω	emu resistance	x 1	= nΩ
coaxial cable: 10^{11} Ω	Gaussian cgs	x 898.76	= GΩ
finger to finger resistance	s/cm	x 898.76	= GΩ
potential: 50 kΩ	International		
integrated circuit diffused	(1948) ohm	x 1.000495	= Ω
resistors: <10 kΩ			

CONVERSIONS for RESISTANCE PER UNIT LENGTH		dimensions: ML/I^2T^3	
Resistivity			
Ohm-metre, Ω.m	statohm.cm	x 8.9876	= GΩ.m
	esu resistance	x 8.9876	= GΩ.m
steel 620 nΩ.m	abohm.cm	x 10	= pΩ.m
copper 17 nΩ.m	emu resistance	x 10	= pΩ.m
nichrome 1000 nΩ.m	microohm-cm	x 10	= nΩ.m
semiconductors 1 μΩ.m to 5 kΩ.m	ohm-cm	x 0.01	= Ω.m
insulators 10 kΩ.m to 10^4 GΩ.m	ohm-circular		
	mil per ft	x 1.662	= nΩ.m

CONVERSIONS for CONDUCTANCE		dimensions: I^2T^3/ML^2	
Electrical conductivity			
Siemens, S = A/V	statmho	x 1.113	= pS
	abmho	x 1	= GS

CONVERSIONS for CONDUCTANCE PER UNIT LENGTH		dimensions: I^2T^3/ML^3	
Specific conductivity, length conductivity			
Siemens per metre, S/m	statmho/cm	x 0.1113	= nS/m
liquid solutions 1 mS/m	abmho/cm	x 10	= TS/m

CONVERSIONS for CONDUCTANCE PER UNIT VOLUME		dimensions: I^2T^3/ML^5	
Volume conductivity			
Siemens per cubic metre, S/m³	statmho/cm³	x 1.113	= μS/m³
	abmho/cm³	x 1000	= TS/m³

CONVERSIONS for CAPACITANCE		dimensions: I^2T^4/ML^2	
Farad, F = C/V	statfarad	x 1.113	= pF
large capacitances to correct power factor 100 μF	esu capacitance	x 1.113	= pF
DC power supplies, rectification 10 mF	abfarad	x 1	= GF
audio amplifier bypass 50 μF	emu capacitance	x 1	= GF
radio frequency bypass 1 nF	Gaussian-cm	x 1.113	= pF
oscilloscope probe-tip capacitance 8 pF	International		
human body capacitance 100 pF	(1948) farad	x 0.999505	= F
two, 3 cm conductors glued side by side 1 pF			
integrated circuit device 5 pF			

CONVERSIONS for MAGNETIC VOLUME SUSCEPTIBILITY		dimensions: dimensionless	
rationalized unit	emu/cm³	x 4 π	= rationalized unit

CONVERSIONS for MAGNETIC MASS SUSCEPTIBILITY		dimensions: L^3/M	
cubic metres per kilogram, m³/kg	emu/g	x 4 πE-3	= m³/kg
	cm³/g	x 4 πE-3	= m³/kg

CONVERSIONS for MAGNETIC FLUX	dimensions: ML^2/IT^2		
Weber, Wb= V.s	esu line	x 0.2998	= kWb
	maxwell	x 0.01	= μWb
iron core magnet for HiFi speaker — *1 mWb*	Gauss-cm²	x 0.01	= μWb
	Maxwell	x 10	= nWb
	unit pole	x 0.12566	= μWb
	kiloline	x 10	= μWb

CONVERSIONS for MAGNETIC FLUX DENSITY	dimensions: M/IT^2		
Magnetic induction, magnetic force vector			
Tesla, T = Wb/m²	esu	x 0.2998	= MT
	gauss	x 0.10	= mT
iron core magnet — *1 T*	gamma	x 1	= nT
air core magnet — *1 mT*	kiloline/in²	x 15.5	= mT
magnetic deflecting yoke of TV	A/m	x 4 πE-7	= T
picture tube provides field of — *2 mT*		x 0.4 π	= μT
earth's magnetic field — *0.1 mT*			
strong bar magnet — *0.2 T*			

CONVERSIONS for INDUCTANCE	dimensions: ML^2/I^2T^2		
Henry, H = Wb/A	stathenry	x 0.8988	= TH
5 cm length of 1 mm diameter wire — *10 nH*	esu inductance	x 0.8988	= TH
for very high radio frequencies — *μH*	Gaussian esu	x 0.8988	= TH
usual radio frequencies — *μH to mH*	abhenry	x 1	= nH
audio frequencies — *H*	emu inductance	x 1	= nH
iron core electromagnet at a	International		
scrap yard: — *10 H*	(1948) henry	x 1.000495	= H
small transformer with multiturns of			
very fine wire: — *5000 H*			

CONVERSIONS for LUMINOUS INTENSITY	dimensions: dimensionless
Light intensity	
candela, cd	
LED (light emitting diode) *= 10 mcd*	

CONVERSIONS for LUMINANCE	dimensions: $1/L^2$		
candela per square metre, cd/m²	candle/ft²	x 10.764	= cd/m²
	candle/in²	x 1.55	= kcd/m²
	ft-lambert	x 3.426	= cd/m²

CONVERSIONS for LUMINEX FLUX	dimensions:
lumen, ℓm = cd.sr	
100 watt bulb emits — *1.7 kℓm*	

CONVERSIONS for LUMINEX FLUX PER UNIT AREA	dimensions: $1/L^2$		
lux = lumens per square metre, ℓm/m²	ft candle	x 10.76391	= ℓm/m²

CONVERSIONS for TIME	dimensions: T		
second, s	minute	x 60	= s
	h	x 3600	= s
	day	x 86400	= s
	a	x 31.55	= Ms

CONVERSIONS for FREQUENCY		dimensions: 1/T
hertz, Hz = 1 cycle per second		
electrical circuit	*60 Hz*	
audio frequency	*kHz*	
radio frequency	*MHz*	

CONVERSIONS for ROTATIONAL FREQUENCY	dimensions: 1/T		
rotational speed, angular velocity			
radians per second, 1/s	r/min	x 2π/60	= rad/s
revolution per minute is commonly used and is accepted by the supplementary practice guide 1978.	rad/s	x 60/2π	= r/min
rpm for motors nominal 1800 r/min (1750 r/min actual)			

CONVERSIONS for PLANE ANGLE	dimensions: dimensionless		
radian, rad	degree	x 0.01745	= rad
circle = 2π *rad*	minute (angle)	x 2.9088E4	= rad
	second (angle)	x 4.848E6	= rad

CONVERSIONS for SOLID ANGLE	dimensions: dimensionless	
steradian, sr	sphere	x
	square degree	x
	hemisphere	x
	spherical right angle	x

SOME PHYSICAL AND ENGINEERING CONSTANTS				
air	23% O_2	76.8% N_2 w/w	π	3.1416
	21% O_2	79% N_2 v/v	e	2.7183
			Avogardo's number, ℕ	6.02217 E23 particles/ mol
molar volume	22.4 L at 0°C and 101.325 kPa		Ideal gas constant, ℝ	8.3143 J/mol.K
std. atmosphere		101.325 kPa		8.3143 Pa.m^3/mol.K
std. gravity acceleration		9.80665 m/s^2	Boltzman constant, **k**	1.38 E-23 J/K
velocity of sound in dry air			Faraday's constant	9.6487 E4 C/equivalent
at 0°C		331.5 m/s	charge on an electron	1.60219 E-19 C
mass density of water at 4°C		0.999973 Mg/m^3	velocity of light, **c**	2.997925 E8 m/s
temperature at absolute zero		-273.15°C	mass of an electron	9.10956 E-31 kg
			mass of a proton	1.67239 E-27 kg
			mass of a neutron	1.6747 E-27 kg

TEMPERATURE CONVERSION TABLES

−459.4 to 0			0 to 100						100 to 1000						1000 to 2000						2000 to 3000					
C		F	C		F	C		F	C		F	C		F	C		F	C		F	C		F	C		F
−273	**−459.4**		−17.8	**0**	32	10.0	**50**	122.0	37.8	**100**	212	260	**500**	932	538	**1000**	1832	816	**1500**	2732	1093	**2000**	3632	1371	**2500**	4532
−268	**−450**		−17.2	**1**	33.8	10.6	**51**	123.8	43	**110**	230	266	**510**	950	543	**1010**	1850	821	**1510**	2750	1099	**2010**	3650	1377	**2510**	4550
−262	**−440**		−16.7	**2**	35.6	11.1	**52**	125.6	49	**120**	248	271	**520**	968	549	**1020**	1868	827	**1520**	2768	1104	**2020**	3668	1382	**2520**	4568
−257	**−430**		−16.1	**3**	37.4	11.7	**53**	127.4	54	**130**	266	277	**530**	986	554	**1030**	1886	832	**1530**	2786	1110	**2030**	3686	1388	**2530**	4586
−251	**−420**		−15.6	**4**	39.2	12.2	**54**	129.2	60	**140**	284	282	**540**	1004	560	**1040**	1904	838	**1540**	2804	1116	**2040**	3704	1393	**2540**	4604
−246	**−410**		−15.0	**5**	41.0	12.8	**55**	131.0	66	**150**	302	288	**550**	1022	566	**1050**	1922	843	**1550**	2822	1121	**2050**	3722	1399	**2550**	4622
−240	**−400**		−14.4	**6**	42.8	13.3	**56**	132.8	71	**160**	320	293	**560**	1040	571	**1060**	1940	849	**1560**	2840	1127	**2060**	3740	1404	**2560**	4640
−234	**−390**		−13.9	**7**	44.6	13.9	**57**	134.6	77	**170**	338	299	**570**	1058	577	**1070**	1958	854	**1570**	2858	1132	**2070**	3758	1410	**2570**	4658
−229	**−380**		−13.3	**8**	46.4	14.4	**58**	136.4	82	**180**	356	304	**580**	1076	582	**1080**	1976	860	**1580**	2876	1138	**2080**	3776	1416	**2580**	4676
−223	**−370**		−12.8	**9**	48.2	15.0	**59**	138.2	88	**190**	374	310	**590**	1094	588	**1090**	1994	866	**1590**	2894	1143	**2090**	3794	1421	**2590**	4694
−218	**−360**		−12.2	**10**	50.0	15.6	**60**	140.0	93	**200**	392	316	**600**	1112	593	**1100**	2012	871	**1600**	2912	1149	**2100**	3812	1427	**2600**	4712
−212	**−350**		−11.7	**11**	51.8	16.1	**61**	141.8	99	**210**	410	321	**610**	1130	599	**1110**	2030	877	**1610**	2930	1154	**2110**	3830	1432	**2610**	4730
−207	**−340**		−11.1	**12**	53.6	16.7	**62**	143.6	100	**212**	413	327	**620**	1148	604	**1120**	2048	882	**1620**	2948	1160	**2120**	3848	1438	**2620**	4748
−201	**−330**		−10.6	**13**	55.4	17.2	**63**	145.4	104	**220**	428	332	**630**	1166	610	**1130**	2066	888	**1630**	2966	1166	**2130**	3866	1443	**2630**	4766
−196	**−320**		−10.0	**14**	57.2	17.8	**64**	147.2	110	**230**	446	338	**640**	1184	616	**1140**	2084	893	**1640**	2984	1171	**2140**	3884	1449	**2640**	4784
−190	**−310**		− 9.44	**15**	59.0	18.3	**65**	149.0	116	**240**	464	343	**650**	1202	621	**1150**	2102	899	**1650**	3002	1177	**2150**	3902	1454	**2650**	4802
−184	**−300**		− 8.89	**16**	60.8	18.9	**66**	150.8	121	**250**	482	349	**660**	1220	627	**1160**	2120	904	**1660**	3020	1182	**2160**	3920	1460	**2660**	4820
−179	**−290**		− 8.33	**17**	62.6	19.4	**67**	152.6	127	**260**	500	354	**670**	1238	632	**1170**	2138	910	**1670**	3038	1188	**2170**	3938	1466	**2670**	4838
−173	**−280**		− 7.78	**18**	64.4	20.0	**68**	154.4	132	**270**	518	360	**680**	1256	638	**1180**	2156	916	**1680**	3056	1193	**2180**	3956	1471	**2680**	4856
−169	**−273**	−459.4	− 7.22	**19**	66.2	20.6	**69**	156.2	138	**280**	536	366	**690**	1274	643	**1190**	2174	921	**1690**	3074	1199	**2190**	3974	1477	**2690**	4874
−168	**−270**	−454	− 6.67	**20**	68.0	21.1	**70**	158.0	143	**290**	554	371	**700**	1292	649	**1200**	2192	927	**1700**	3092	1204	**2200**	3992	1482	**2700**	4892
−162	**−260**	−436	− 6.11	**21**	69.8	21.7	**71**	159.8	149	**300**	572	377	**710**	1310	654	**1210**	2210	932	**1710**	3110	1210	**2210**	4010	1488	**2710**	4910
−157	**−250**	−418	− 5.56	**22**	71.6	22.2	**72**	161.6	154	**310**	590	382	**720**	1328	660	**1220**	2228	938	**1720**	3128	1216	**2220**	4028	1493	**2720**	4928
−151	**−240**	−400	− 5.00	**23**	73.4	22.8	**73**	163.4	160	**320**	608	388	**730**	1346	666	**1230**	2246	943	**1730**	3146	1221	**2230**	4046	1499	**2730**	4946
−146	**−230**	−382	− 4.44	**24**	75.2	23.3	**74**	165.2	166	**330**	626	393	**740**	1364	671	**1240**	2264	949	**1740**	3164	1227	**2240**	4064	1504	**2740**	4964
−140	**−220**	−364	− 3.89	**25**	77.0	23.9	**75**	167.0	171	**340**	644	399	**750**	1382	677	**1250**	2282	954	**1750**	3182	1232	**2250**	4082	1510	**2750**	4982
−134	**−210**	−346	− 3.33	**26**	78.8	24.4	**76**	168.8	177	**350**	662	404	**760**	1400	682	**1260**	2300	960	**1760**	3200	1238	**2260**	4100	1516	**2760**	5000
−129	**−200**	−328	− 2.78	**27**	80.6	25.0	**77**	170.6	182	**360**	680	410	**770**	1418	688	**1270**	2318	966	**1770**	3218	1243	**2270**	4118	1521	**2770**	5018
−123	**−190**	−310	− 2.22	**28**	82.4	25.6	**78**	172.4	188	**370**	698	416	**780**	1436	693	**1280**	2336	971	**1780**	3236	1249	**2280**	4136	1527	**2780**	5036
−118	**−180**	−292	− 1.67	**29**	84.2	26.1	**79**	174.2	193	**380**	716	421	**790**	1454	699	**1290**	2354	977	**1790**	3254	1254	**2290**	4154	1532	**2790**	5054
−112	**−170**	−274	− 1.11	**30**	86.0	26.7	**80**	176.0	199	**390**	734	427	**800**	1472	704	**1300**	2372	982	**1800**	3272	1260	**2300**	4172	1538	**2800**	5072
−107	**−160**	−256	− 0.56	**31**	87.8	27.2	**81**	177.8	204	**400**	752	432	**810**	1490	710	**1310**	2390	988	**1810**	3290	1266	**2310**	5090	1543	**2810**	5090
−101	**−150**	−238	0	**32**	89.6	27.8	**82**	179.6	210	**410**	770	438	**820**	1508	716	**1320**	2408	993	**1820**	3308	1271	**2320**	4208	1549	**2820**	5108
− 95.6	**−140**	−220	0.56	**33**	91.4	28.3	**83**	181.4	216	**420**	788	443	**830**	1526	721	**1330**	2426	999	**1830**	3326	1277	**2330**	4226	1554	**2830**	5126
− 90.0	**−130**	−202	1.11	**34**	93.2	28.9	**84**	183.2	221	**430**	806	449	**840**	1544	727	**1340**	2444	1004	**1840**	3344	1282	**2340**	4244	1560	**2840**	5144
− 84.4	**−120**	−184	1.67	**35**	95.0	29.4	**85**	185.0	227	**440**	824	454	**850**	1562	732	**1350**	2462	1010	**1850**	3362	1288	**2350**	4262	1566	**2850**	5162
− 78.9	**−110**	−166	2.22	**36**	96.8	30.0	**86**	186.8	232	**450**	842	460	**860**	1580	738	**1360**	2480	1016	**1860**	3380	1293	**2360**	4280	1571	**2860**	5180
− 73.3	**−100**	−148	2.78	**37**	98.6	30.6	**87**	188.6	238	**460**	860	466	**870**	1598	743	**1370**	2498	1021	**1870**	3398	1299	**2370**	4298	1577	**2870**	5198
− 67.8	**− 90**	−130	3.33	**38**	100.4	31.1	**88**	190.4	243	**470**	878	471	**880**	1616	749	**1380**	2516	1027	**1880**	3416	1304	**2380**	4316	1582	**2880**	5216
− 62.2	**− 80**	−112	3.89	**39**	102.2	31.7	**89**	192.2	249	**480**	896	477	**890**	1634	754	**1390**	2534	1032	**1890**	3434	1310	**2390**	4334	1588	**2890**	5234
− 56.7	**− 70**	− 94	4.44	**40**	104.0	32.2	**90**	194.0	254	**490**	914	482	**900**	1652	760	**1400**	2552	1038	**1900**	3452	1316	**2400**	4352	1593	**2900**	5252
− 51.1	**− 60**	− 76	5.00	**41**	105.8	32.8	**91**	195.8				488	**910**	1670	766	**1410**	2570	1043	**1910**	3470	1321	**2410**	4370	1599	**2910**	5270
− 45.6	**− 50**	− 58	5.56	**42**	107.6	33.3	**92**	197.6				493	**920**	1688	771	**1420**	2588	1049	**1920**	3488	1327	**2420**	4388	1604	**2920**	5288
− 40.0	**− 40**	− 40	6.11	**43**	109.4	33.9	**93**	199.4				499	**930**	1706	777	**1430**	2606	1054	**1930**	3506	1332	**2430**	4406	1610	**2930**	5306
− 34.4	**− 30**	− 22	6.67	**44**	111.2	34.4	**94**	201.2				504	**940**	1724	782	**1440**	2624	1060	**1940**	3524	1338	**2440**	4424	1616	**2940**	5324
− 28.9	**− 20**	− 4	7.22	**45**	113.0	35.0	**95**	203.0				510	**950**	1742	788	**1450**	2642	1066	**1950**	3542	1343	**2450**	4442	1621	**2950**	5342
− 23.3	**− 10**	14	7.78	**46**	114.8	35.6	**96**	204.8				516	**960**	1760	793	**1460**	2660	1071	**1960**	3560	1349	**2460**	4460	1627	**2960**	5360
− 17.8	**0**	32	8.33	**47**	116.6	36.1	**97**	206.6				521	**970**	1778	799	**1470**	2678	1077	**1970**	3578	1354	**2470**	4478	1632	**2970**	5378
			8.89	**48**	118.4	36.7	**98**	208.4				527	**980**	1796	804	**1480**	2696	1082	**1980**	3596	1360	**2480**	4496	1638	**2980**	5596
			9.44	**49**	120.2	37.2	**99**	210.2				532	**990**	1814	810	**1490**	2714	1088	**1990**	3614	1366	**2490**	4514	1643	**2990**	5414
						37.8	**100**	212.0				538	**1000**	1832				1093	**2000**	3632				1649	**3000**	5432

NOTE—The numbers in **Bold Face Type** refer to the temperature either in degrees Centigrade or Fahrenheit which it is desired to convert into the other scale. If converting Fahrenheit degrees to Centigrade degrees, the equivalent temperature will be found in the left column, while if converting from degrees Centigrade to degrees Fahrenheit, the answer will be found in the column on the right. (Reprinted courtesy of Scovill Fasteners, Inc.)

PART B: CONVERTING CONCENTRATIONS

One of the challenges in selecting and sizing equipment is to appropriate mass balance over the unit. Mass balance, by definition, means that the **mass** is needed in the balancing calculation. However, sometimes we are given concentrations and volumetric flowrates. Indeed, the design procedures for some equipment are centered around concentrations and volumetric flowrates.

MASS FRACTION TO MASS CONCENTRATION FOR SOLUTIONS

To convert concentrations to mass requires a knowledge of the density of the mixture. As an *approximation* we might assume that the volumes are additive in the mixture. That is, if a mixture is 30% w/w of a component of density 1600 kg/m^3 and 60% water, then the concentration is, if the volumes are additive:

$$\frac{0.30}{\frac{0.3}{1600}+\frac{0.70}{1000}} = 340 \text{ g/L}$$

Figure B–1 shows the results of these calculations **for aqueous systems,** based on the additive volume approximation. This is reasonable for many systems. Two features are included in Figure B–1: the usual main graph and the exceptions.

For the *usual* systems the error is less than 10%. The pure component density for the solute is given on the RHS ordinate. The wt% is given on the abscissa. The resulting "concentration" is given on the LHS ordinate. For example, for 50% w/w ethanol in water, what is the concentration of ethanol at 20°C? The procedure is:

1. estimate the density of the solute. Values for some solutes are given at the top and right-hand side of this graph for 25°. (Data for other solutes and for other temperatures can be estimated from Part Db). In the example problem, since the temperature is close to 25°C, we can select the appropriate line for ethanol directly, at 0.80.
2. enter the graph on the abscissa at the desired mass fraction. In this example this is 0.50.
3. rise vertically to intersect the density line chosen in step 1 and read the answer on the LHS abscissa (in the central, ideal line). In this example, the value is about 450 g/L. (The value from reference books is 456.92 g/L.)

For the *exceptions,* such as sulfuric, hydrochloric, and hydrofluoric acids, the additive volume approximation gives > 10% error. For these solutes, we follow the same procedure. We rise vertically from the specified mass fraction to the line for the "ideal" pure component density. This is given on the RHS for 25°C for these solutes. However, when projecting the intersection horizontally toward the LHS ordinate, instead of stopping at the "central, ideal" ordinate, we continue further to the nonideal concentration ordinate that is specific to that species.

MASS % TO VOLUME % FOR AQUEOUS SLURRIES OF SOLIDS

$$\text{vol\% solid} = \frac{\text{wt\%}/\rho \text{ solids}}{\text{wt\%}/\rho \text{ solids} + ((100 - \text{wt\%})/\rho \text{ liquid})} \tag{B-1}$$

The volume % solids in a water slurry can be converted to the mass % solids by the following relationship:

$$\text{mass \% solid} = \frac{\text{vol\%} \times \text{sp. gr. solid}}{\text{vol \%} \times \text{sp. gr. solid} + (100 - \text{vol \%}) \times 1.}$$

A plot of Equation B–1 is given in Figure B–2 for water as the solvent. For these calculations water was assigned a density of 1 Mg/m^3.

MOL FRACTION TO MOLAR CONCENTRATION & MOLALITY FOR LIQUID SOLUTIONS

Molar concentration is expressed as mol fractions, molar concentration (Molarity), and sometimes as Molality.

Mol fraction is the ratio of the mols of the target species "2," n_2, to the total number of mols in the solution. For a liquid, this is commonly represented by x_2.

The *molar concentration,* c_2 and sometimes referred to as M or molarity, is the number of moles of solute "2," n_2, per liter of solution.

$$\tilde{c}_2 = \frac{n_2}{\text{1L solution}} = \text{mol/L} \tag{B-2}$$

Molar concentration depends on the *temperature.*

If the molar volume of the solute "2" is V_2 and that of the solvent is V_1, then the relationship between mol fraction and molar concentration (assuming additivity of volumes) is

$$\tilde{x}_2 = \frac{\tilde{c}_2\tilde{V}_1}{\tilde{c}_1\tilde{V}_2 + (1000 - \tilde{c}_2\tilde{V}_2)} \tag{B-3}$$

For small concentrations, this becomes:

$$\tilde{x}_2 = 0.001\, \tilde{c}_2 \left(\frac{M_1}{\rho_1}\right) \tag{B-4}$$

Less frequently used is molality, m_2. This temperature independent term is defined as the number of moles of solute "2" per 1000 g of "pure" solvent. Thus,

$$m_2 = \frac{n_2}{\text{1000g solvent}} \tag{B-5}$$

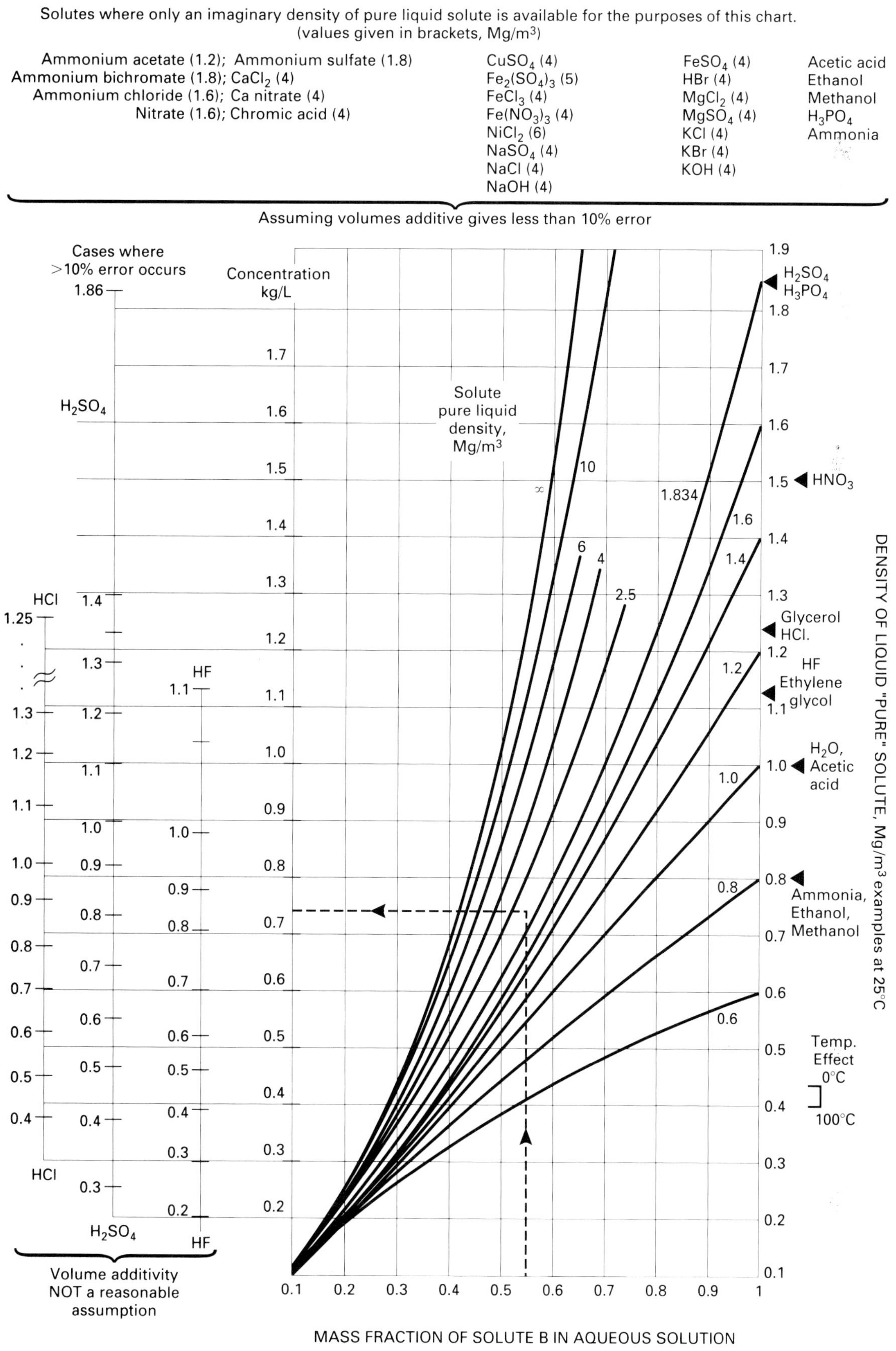

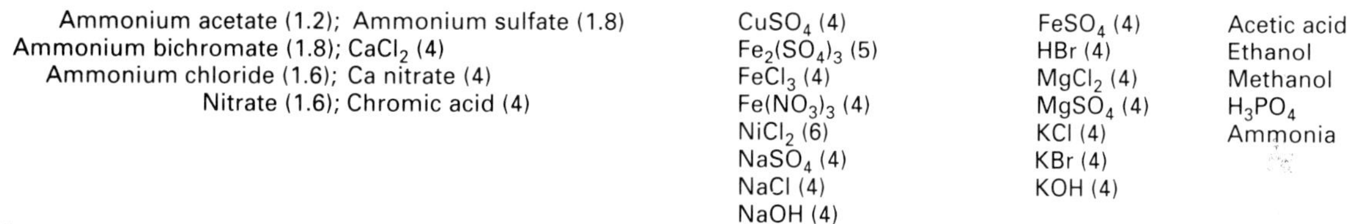

Solutes where only an imaginary density of pure liquid solute is available for the purposes of this chart.
(values given in brackets, Mg/m3)
Ammonium acetate (1.2); Ammonium sulfate (1.8)
Ammonium bichromate (1.8); CaCl2 (4)
Ammonium chloride (1.6); Ca nitrate (4)
Nitrate (1.6); Chromic acid (4)
CuSO4 (4)
Fe2(SO4)3 (5)
FeCl3 (4)
Fe(NO3)3 (4)
NiCl2 (6)
NaSO4 (4)
NaCl (4)
NaOH (4)
FeSO4 (4)
HBr (4)
MgCl2 (4)
MgSO4 (4)
KCl (4)
KBr (4)
KOH (4)
Acetic acid
Ethanol
Methanol
H3PO4
Ammonia
Assuming volumes additive gives less than 10% error
Cases where
>10% error occurs
Concentration
kg/L
Solute
pure liquid
density,
Mg/m3
H2SO4
HCl
HF
H2SO4
H3PO4
HNO3
Glycerol
HCl.
HF
Ethylene
glycol
H2O,
Acetic
acid
Ammonia,
Ethanol,
Methanol
Temp.
Effect
0°C
100°C
Volume additivity
NOT a reasonable
assumption
DENSITY OF LIQUID "PURE" SOLUTE, Mg/m3 examples at 25°C
MASS FRACTION OF SOLUTE B IN AQUEOUS SOLUTION

Figure B–1 Relating the Mass Fraction to Mass Concentration for Solutions

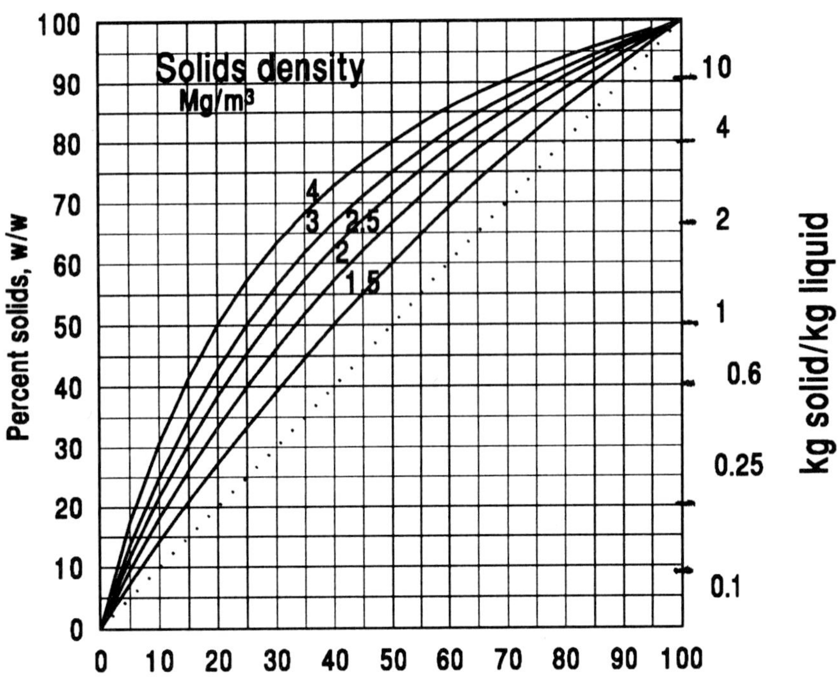

Figure B–2 Relating Volume % to Mass % for Aqueous Slurries of Solids

The relationship between mol fraction and molality is given by:

$$\tilde{x}_2 = \frac{0.001\ m_2M_1}{1 + 0.001\ m_2M_1} \qquad \text{(B-6)}$$

For small concentrations this becomes the same as Equation B–4:

$$\tilde{x}_2 = 0.001\ m_2M_1 \qquad \text{(B-7)}$$

Figure B–3 shows mol fraction x_2 plotted against the molar concentration c_2. The solid black lines show Equation B–6 for molality and, for small concentrations, Equations B–4 and B–7. Thus, different lines are for different values of the molar volume, V_1, for the solvent.

For higher molar concentrations, greater than about 0.5 M, the molar concentration-mol fraction relationship depends on the molar volume of the solute, V_2, as one would expect from Equation B–3. When the molar volumes of the solute and the solvent are equal, the plot is a straight line of slope 1. When the molar volumes are different, the variation is as shown in Figure B–3 (for the *solvent molar volumes* of 18 and 300) and in Figure B–4 (for solvent molar volumes of 50 and 200). When the solute molar volume is *greater* than that of the solvent, the line curves upward; when it is less, the line curves downward.

Example, at 25°C what is the mol fraction of MEA in a 5 M solution of monoethanolamine? At 25°C, the molar volume of water is 18 and that of MEA is 60 (from Part Da). Select the *solvent* line: 18. Enter the graph at the abscissa of 5 M or 5 mol/L, rise to the solvent line corresponding to 18, adjust for the solute correction (continue moving vertically to the solute line of 60) and then read off the mol fraction from the ordinate. In this example, the answer is about 0.12 MEA.

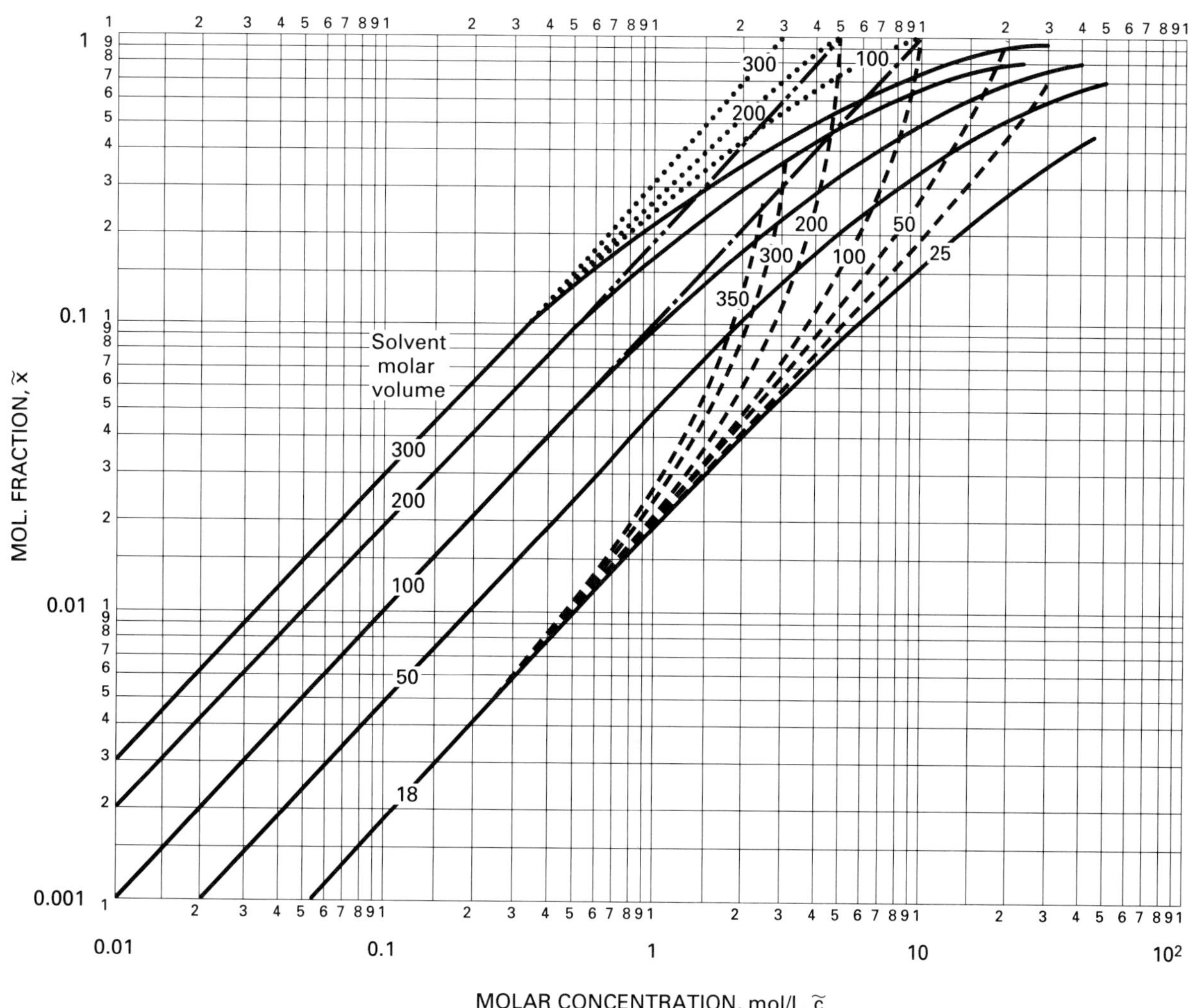

Figure B–3 Relating Mol Fraction to Molar Concentration and Molality for Liquid Solutions

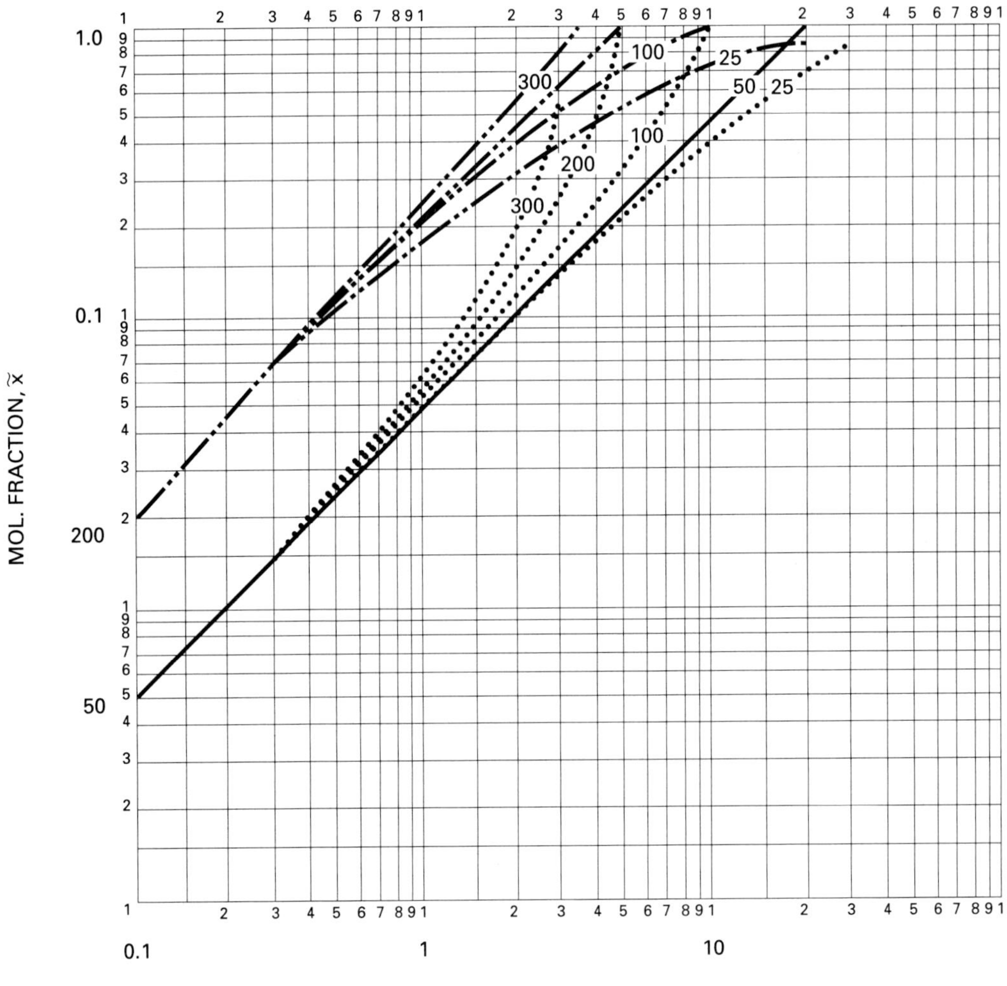

Figure B–4 Relating Mol Fraction to Molar Concentration for Liquid Solutions

PART C: PROPERTIES OF MATERIALS

We cannot select or size equipment unless we know the pertinent properties of the gases, liquids, and solids being processed. Tables C–1 through C–3 focus on properties needed to characterize particles and to select and size equipment to **transport** solids, slurries, dusts, and powders. Table C–4 gives data for particles that is useful in selecting options to separate systems containing solids and solids from other solids.

Table C-1 provides the Codes used, in this **Data** book and in the companion book on **Process Design and Engineering Practice,** for characterizing particles. Table C-2 compares the different Mesh sizes used to characterize particle size, and relates that to the size code.

For transportation, Table C-3 summarizes the usual size, flowability, abrasiveness, special characteristics and density of various solids. Table C-3 lists the compound name, characteristics and *bulk* density of the powder form. Then listed are the pertinent sizing data needed for different alternatives. For a screw conveyor, the key information is the trough loading capacity (expressed as a percentage) and the power factor (a dimensionless number needed to estimate the power required). For a belt conveyor, the use-

TABLE C-1: Code for characterizing solids

Size		
Code	Meaning: diameter	Terminology
α	<250 Mesh	silt
A	<100 Mesh	very fine
B	<0.3 cm	fine
C	<1.3 cm	granular
D	1.3 to 1.5 cm	pebble
δ	5 to 10 cm	small cobble
Δ	10 to 15 cm	medium cobble
E		fibrous and stringy
Flowability		
Code	Meaning: angle of repose	Terminology
1	<15°	flushing
2	15 to 30°	free flowing
3	30 to 45°	average flowability
4	45 to 60°	sluggish
5	60 to 90°	sticky
Abrasiveness		
Code	Meaning: moh hardness	Terminology
6	1	non-abrasive
7	2 to 3	mildly abrasive
8	3 to 5	moderately abrasive
9	>5	very abrasive
Corrosivity		
Code	Meaning: pH	Terminology
F	>7	non-corrosive
G	5 to 7	mildly corrosive
H	1.5 to 5	very corrosive

Table C-1: (Continued)

Handling Characteristics		
Code	Meaning	
J	fluidizes	
K	cakes, builds up, hardens	
L	light and fluffy	
M	dusty	
N	sticky	
O	hygroscopic	
P	becomes plastic	
Q	packs under pressure	
R	interlocks, mats or agglomerates	
Safety-hazard		
Code	Meaning	
S	generates static electricity	
T	toxic fumes given off	
U	flammability	
V	explosive dust	
Product considerations		
Code	Meaning	
W	prevent contamination	
X	prevent degradation	
Y	prevent decomposition	

ful information is the angle of repose and the maximum belt incline that can be used. (These are both expressed in degrees relative to the horizontal.) Bucket elevator and feeders should not be used with certain materials. Data are so indicated in columns 8 and 9. For hydraulic conveying, the key data are the solid (not powder) density, the usual design loading expressed is kg solid/kg water and the energy required per Mg/h of solids conveyed per 1000 m of distance. For pneumatic conveying, a wide variety of data have been published. These are listed in columns 13 to 20. First, for vertical conveying, the minimum gas velocity (m/s) and the maximum solids loading (kg solid/kg gas) are given. Then, the analogous data for horizontal conveying. Usually the horizontal conditions are the constraining conditions. In columns 17 to 20, the usual design gas velocity (m/s) and the usual solids loading (kg solid/kg gas) are given for vacuum and for pressure conveying respectively operating at a distance of 120 m equivalent pipe length.

Table C-4 provides data for selecting and sizing physical separations.

For separating species starting from a heterogeneous phase, what are the key properties for us to consider?

- Size of the particulate or dispersed phase. Most of the initial selection procedures need to know the size of the particulate. Sometimes we work with the size that occurs naturally: the size of dust emitted, the size of an emulsion generated, the size of a crystal grown.
- The pure solid density is a key to understanding how easy it will be to use gravity and centrifugal forces to effect the separation. This density is **not** to be confused with the "bulk" density reported in Table C–2. The latter is the density of the particles and air in the interstices.
- The hardness of the solid, the "Work Index," and the Abrasion Index that give an indication of the amount of energy required to reduce the size of the particles. We often want to change the "natural" size of the particles. Sometimes we wish to **reduce** the size of the initial particle.

TABLE C-2. Relating US standard mesh size to particle size and comparison with other sieves

Code	US Std.		US Tyler		German Std.		French Std.		British Std.		1 MM	
	Mesh size	nom. open, μm	Mesh size	nom. open, μm	Mesh size	nom. open, μm	Mesh size	nom. open, μm	Mesh size	nom. open, μm	Mesh size	nom. open, μm
a	325	44	325	43								
	270	53	270	53					300	52		
	230	63	230	61	100	60	250	58	240	66	200	63
	200	74	200	74	80	75	200	75	200	75	150	86
	170	88	170	88	70	88			170	90		
	140	105	150	104	60	100	150	100	150	102	120	108
	120	125	115	125	50	120			120	125	100	128
A					45	135					90	138
	100	149	100	147	40	150	100	150	100	150	80	157
	80	177	80	175					85	178	70	180
	70	210	65	208	30	200			72	211	60	210
	60	250	60	248	24	250			60	253	50	255
	50	297	48	295	20	300	50	320	44	353		
	40	420	35	417					36	420	30	420
	30	590	28	589					30	500		
B	20	840	20	833					22	699		
	18	1000	16	991					16	1000		
	16	1190	14	1168								
	14	1410	12	1397								
	12	1680	10	1651					10	1650		
	10	2000	9	1981								
	8	2380	8	2362								
C	4	4760	4	4699								

- The condition to create a "zero point of charge" on the dispersed solid in aqueous solution is important for selecting "froth flotation" (as an option to separate solid from solids) and to increase the size of particles.
- The electrostatic behavior of solids in air and the magnetic properties of solids are important in determining if such differences can be exploited.

Consider each in turn.

The name of the solid (and an indication of the formula) are given in the first two columns of Table C–4. The chemical index should be consulted first because the entry in Table C–4 is mainly by mineral name. Thus, barium sulfate is located under "synthetic" Barite.

The mass density is relatively independent of pressure and temperature. Data from a wide variety of sources are summarized here in column 3.

The hardness of a mineral is often given in units of Mohs' hardness scale with talc assigned a value of 1 and diamond a value of 10. The relationship between Mohs scale and the Brinell and Vicker's scale is given in Figure 2–43. A measure of the amount of energy required to crush a solid is given in the Work Index. This is defined as the amount of power (in kWh) required to crush an "infinitely" large rock so that 80% is smaller than 100 μm (or such that 67% is –200 Mesh). The units are kWh $(\mu m)^{0.5}$/Mg.

The pH condition when the surface of a dispersed material has zero charge is given in column 6. Here is the background. Usually, when a solid (or oil) is dispersed in water, either H^+ or OH^- ions adsorb preferentially to the surface. If an excess of OH^- ions are adsorbed, the net surface charge is **negative** and vice versa. It is wise to always assume that such a surface charge exists in any aqueous system. These charge effects become significant when the size of the dispersed particles is –200 Mesh or less than 75 μm and become more important as the size drops below 1 μm. Thus, for sizes where "froth flotation" might be an option, for sizes where deep bed filters might be used, or for conditions where we want to **increase** the size, some idea of the magnitude and sign of the surface charge is important. The zpc (or the point when the surface charge is zero) is the pH (or bulk concentration of the ions causing the

Table C-3 Solids: Particulate bulk density, characteristics and conveyability.

Material	Size/code	Bulk Den Mg/ cu metre	Screw %	conve factor	Belt co repose o	tilt o	Bu el. ?	Fe ?	Bl tk ?	Air sld ?	Hydraul convey kg/kg	Pneumatic vert m/s	kg/kg	Conveying horiz m/s	kg/kg	vacuum m/s	kg/kg	lo press m/s	kg/kg
Adipic acid:	A37O	0.72	30	0.8															
Alfalfa meal:	B47L	0.22-0.35	30	0.6												33	3.3	21	7
Alfalfa pellets:	C27	0.67	45	0.6															
Alfalfa seed:	B17V	0.16-0.24	45	0.5															
Almonds, broken:	C37X	0.43-0.48	30	0.9															
Almonds, whole & shelled:	C37X	0.45-0.48	30	0.9															
Alumina:	B29JL	0.88-1	15	1.8	22	10								35		32	3.3	19	7
Alumina, floury:	A49JLN				>60									24-30					
Alumina, fines:	A29JL	0.56	15	1.6			N	-	-	-		1.5	380	7.6	75				
Alumina sized or briquettes:	D39	1.04	15	2															
Aluminate hydroxide (gel):	B37	0.72	30	1.7															
Aluminum chips, dry:	E47R	0.11-0.24	30	1.2															
Aluminum chips, oily:	E47R	0.11-0.24	30	0.8															
Aluminum hydrate, course:	C37	0.2-0.32	30	1.4	34	24													
Aluminum hydrate, fine:	A37	0.88-1.2														32	3.3	19	7
Aluminum oxide:	A19J	0.96-1.92	15	1.8															
Aluminum silicate (Andalusite):	C37H	0.79	30	0.8															
Aluminum sulfate (Alum):	C27	0.72-0.93	45	1	32	15													
lumpy	B27	0.80-0.96	45	1.4												33	2.8	20	6
pulverized	B37O	0.72-0.8	30	0.6												33	2.8	20	6
fine	A	0.67														33	2.8	20	6
Ammonium chloride:	A47HKT	0.72-0.83	30	0.7															
Ammonium nitrate:	A37VGO	0.72-1	30	1.3															
Ammonium nitrate, prilled:	C27VGO	0.72																	
Ammonium sulfate:	C37KONG	0.72-0.93	30	1															
Antimony powder:	A37		30	1.6															
Apple pomace, dry:	C47L	0.24	30	1															
Arsenic oxide (Arsenolite):	A37T	1.6-1.92	30																
Arsenic, pulverized:	A27T	0.48	45	0.8												38			
Asbestos rock, ore:	D39T	1.30	15	1.2															
Asbestos (Amianthus):	E48	0.32-0.64	30A	0.8															
Asbestos, shredded:	E48QL	0.32-0.64	30A	1			N	-	-	-						low			
Ash, black ground:	B37	1.68	30	2	27	15													
	A	0.72										1.5	380	4.5	130				
Ashes, coal dry	C48GL	0.56-0.72	30A	3	40	27													
Ashes, coal dry:	D48G	0.56-0.64	30A	2.5															
Ashes, coal wet:	C48G	0.72-0.8	30A	3	50	38													
Ashes, coal wet:	D48G	0.72-0.8	30A	4															
Asphalt, crushed	C47	0.72	30	2															
Bagasse:	E47RTQL	0.11-0.16	30	1.5															
Bakelite, fine:	B27	0.48-0.72	45	1.4															
Baking Powder:	A37	0.64-0.72	30	0.6															
Barite (barium sulfate):	D38	1.92-2.88	30A	2.6															
Barite powder:	A37Q	1.92-2.88	30	2								4.5	380	7.6	130				
Barium carbonate:	A47T	1.15	30	1.6	57														
Bark wood refuse:	E47LRG	0.16-0.32	30	2															
Barley, fine ground:	B37	0.54-0.61	30	0.4															
Barley, malted:	C37	0.5	30	0.4															
Barley, meal:	C37	0.45	30	0.4															
Barley, whole:	B27V	0.58-0.77	45	0.5	48	35										Y		Y	
Basalt:	B29	1.28-1.68	15	1.8															
Bauxite, dry ground:	A27	1.44										1.5	400	7.6	104				
Bauxite, dry, ground:	B27	1.09	45	1.8	35	23													
Bauxite, crushed:	D38	1.2-1.36	30A	2.5	31	17													

Beans, castor, meal:	B37	0.56-0.64	30	0.8															
Beans, castor, whole shredded:	C17	0.58	45	0.5															
Beans, Navy, dry:	C17	0.77	45	0.5															
Beans, Navy, steeped:	C27	0.96	45	0.8															
Bentonite, crude:	D47Q	0.54-0.64	30	1.2															
Bentonite, -100 Mesh:	A27JLQ	0.8-0.96	45	0.7								1.5	380	9.1	130	Y		Y	
Blood, dried:	D47O	0.56-0.72	30	2															
Blood, ground dried:	A37O	0.48	30	1															
Bone ash (Tricalcium phosphate):	A47	0.64-0.8	30	1.6															
Boneblack:	A27L	0.32-0.4	45	1.5															
Bonechar:	B37	0.43-0.64	30	1.6															
Bonemeal:	B37	0.8-0.96	30	1.7												33	3.3	21	7
Bones, whole:	E47R	0.56-0.8	30	3															
Bones, crushed:	D47	0.56-0.8	30	2															
Bones, ground:	B37	0.8	30	1.7															
Borate of lime:	A37	0.96	30	0.6															
Borax, fine:	B27G	0.72-0.88	45	0.7															
Borax, fine:	A27G										0.35					36	2.7	28	3.3
Borax, screenings:	C37	0.88-0.96	30	1.5															
Borax, lumps:	D37	0.88-0.96	30	1.8															
Borax, lumps:	D37	0.96-1.12	30	2															
Boric acid, fine:	B27G	0.88	45	0.8												Y		Y	
Boron:	A39	1.2	15	1															
Bran, rice,rye,wheat:	B37LV	0.26-0.32	30	0.5	S														
Braunite (Manganese oxide):	A38	1.92	30A	2															
Bread crumbs:	B37WX	0.32-0.4	30	0.6															
Brewer's grain, spent & dry:	C47	0.22-0.48	30	0.5															
Brewer's grain, spent & wet:	C47G	0.88-0.96	30	0.8															
Brick, ground:	B39	1.6-1.92	15	2.2															
Bronze chips:	B47	0.48-0.8	30	2															
Buckwheat:	B27V	0.59-0.67	45	0.4	25	13													
Calcine, flour:	A37	1.2-1.4	30	0.7															
Calcium carbide:	D27V	1.12-1.45	30	2															
Calcium carbonate: Calcite	A	0.4-0.48														33	3.2	20	6.7
Calcium silicide	U	1.92			N		N	-	-	-	_								
Calcium sulfate, hydrated:(gypsum)																			
calcined powdered:	A37O	0.96-1.28	30	2.0			-	-	-	Y									
Gypsum/ calcined:	B37O	0.88-0.96	30	1.6	40	27													
Gypsum/ ground:	C37O																		use v
Gypsum/ raw -2.5 cm:	D27	1.12-1.28	30	2.0	30	15													
Carbon/ activated: corrosive	AGUJ	0.32					-	N	-	-									
/ activated light:	AGUJ	0.16																	
/ activated:	AGUJ	0.24					-	-	-	-	0.48								
Carbon black/powder	A17LJQU	0.06-0.1			N		N	-	-	-									
Carbon black/ light:	QU	0.24			N		N	-	-	-									
Carbon black/ pelleted:	B17JQUX	0.32-0.4			N		N	-	-	-									
Carborundum/ -75 mm:	D29	1.60	15	3															
Casein	B37O	0.58	30	1.6															
Cashew nuts:	C47	0.5-0.59	30	0.7															
Cast iron/ chips:	C47	2-3.2	30	4															
Catalyst																			
Cellulose acetate/ fine & flake		0.25-0.35														30	3.3	17	7
Cement (portland)	A28JOQ	1.12-1.5	30A	1.4	39	28	-	-	Y	Y		1.5	800	7.6	130				
Cement/ Portland/ aerated	A18J	0.96-1.2	30A	1.4															
Cement/ mortar:	B37X	2.38	30	3.0															
Cement clinker:	D38	1.2-1.5	30A	1.8	33	20													
Cerrusite: (Lead Carbonate)	A37T	3.85-4.2	30	1															
Chalk/ crushed	D27	1.2-1.5	30	1.9															
Chalk/ pulverized	A27JLQ	1.07-1.2	45	1.4															

continued

Table C-3 Solids: Particulate bulk density, characteristics and conveyability (continued).

Material	Size/code	Bulk Den Mg/ cu metre	Screw %	conve factor	Belt co repose o	tilt o	Bu el. ?	Fe ?	Bl tk ?	Air sld ?	Hydraul convey kg/kg	Pneumatic vert m/s	kg/kg	Conveying horiz m/s	kg/kg	vacuum m/s	kg/kg	lo press m/s	kg/kg
Charcoal/ ground	A47	0.29-0.45	30	1.2															
/ granular:	D47X	0.29-0.45	30	1.4	35	25													
Chocolate/ cake pressed:	D27	0.64-0.72	30	1.5															
Chrome ore:	D38	2-2.25	30A	2.5															
Chromite:	7KTQ	1.9																	
Chromium:/	9TV	2.53					N	-	N	-		N		N		N		N	
Cinders/coal:	D38G	0.64-0.67	30A	1.8	35	23													
Cinders/ blast furnace:	D38G	0.91	30A	1.9	35	23													
Citric acid:	GJOW	0.72	N		N		N	-	-	-									
Clay (see also Bentonite/ Diatomaceous earth/ Fuller's earth/ Kaolin and Marl)																			
Clay/ insecticide grade:	OJ	0.19-0.22																	
Clay (airfloat)	OQJ	0.48														31	3.2	15	7
Clay (air float/ rubber grade)		0.5-0.56																	
Clay/ water washed:		0.68-0.8														35	2.8	18	6.35
Clay/ blended for tile:		0.72			45	31													
Clay/ spray dried:		0.96														33	3	17	6.7
Clay/ ceramic and ball:	A37W	0.8-1.2	30	1.5															
Clay/ heavy	K7O	1.12																	
Clay/ lumps	7O	1.2																	
Clay/ dry loose lump:	D37	0.96-1.2	30	1.8	35	21													
Clay/ ground/					35	22													
Clay/ grey and granular:	C				35	20													
Clay/ calcined:	B38	1.3-1.6	30A	2.4															
Clay/ brick/ dry fines:	C38	1.6-1.9	30A	2.0															
Clover seed:	B27V	0.77			28	15													
Coal, anthracite	7	0.64-0.96					-	-	Y	Y	_							Y	
Coal/ anthracite/ broken and loose:		0.87			22	8													
Coal/ anthracite/ chestnut/		0.74			22	8													
Coal/ bituminous	6	0.64-0.96																	
bituminous/ dry-6.5 mm:	C	0.67			29	15													
bituminous/ wet-6.5mm:	CKVU	0.80	N		40	25													
bituminous/ very wet/	C	0.88			33	20													
bituminous/ sized wet/dry		0.72			27	14													
Coal/ lignite:	D37G	0.6-0.72	30	1.0								9	19	12	12				
Coal/ powder	GVUO	0.48										1.5	250	4.5	91				
Cocoa	KVOWQ	0.48	N		N		N	-	-	-									
Cocoa beans							-	-	N	-	_	N				N		N	
Cocoa powder:	A	0.40														Y		Y	
Cocoanuts/ shredded/		0.40			27	15													
Coffee beans/ green		0.67			25	10										22	7.4	13	12
Coffee/ steel cut/		0.45			23	10													
Coke:	A	0.4-0.56														Y		Y	
coke	B9	0.4-0.56														Y		Y	
petroleum	B9	0.56					-	-	-	Y	_					Y		Y	
Copper	Vu	2.96																	
Copper sulfate/ ground	GT	1.14-1.2	31	17															
Copra:	DR	0.35-0.4														Y		Y	
Copra: medium size		0.53	20	9															
Copra meal/ ground/		0.64			39	25													
Copra/ expeller cake/																			
ground:		0.51			30	16													
chopped:		0.46			20	8													

Corn/ ear:	E37	0.90	30	--														
Corn/ germ:	B37LW	0.34	30	0.4														
Corn/grits:	B37W	0.64-0.72																
Corn/ ground:(sugar):	B37VOW	0.48-0.56	30(N)	1.0	N		N	-	-	-					30	3.8	21	8.3
Corn seed:	C27WX	0.72	45	0.4														
Corn shelled:	C27	0.72-0.9	45	0.4	21	7									31	4.8	17	10
Cornmeal:	B37VOW	0.51-0.66	30(N)	0.5	35N	22	N	-	-	-								
Corn oil cake:	D47Y	0.40	30	0.6														
Cottonseed dry not delinted:	C47Q	0.29-0.4	30	0.9											N		N	
Cottonseed/dry delinted:	C27Q	0.35-0.64	45	0.6	29	19									Y		Y	
Cottonseed hulls:	B37L	0.19	30	0.9														
Cottonseed/ cake crushed:	C47Y	0.64-0.72	30	1.0														
Cottonseed/ cake lumpy:	D47Y	0.64-0.72	30	1.0														
Cottonseed flakes:	C37ML	0.32-0.4	30	0.8														
Cottonseed meal expeller:	B47Y	0.4-0.48	30	0.5											Y		Y	
Cottonseed meal extracted:	B47Y	0.56-0.64	30	0.5	35	22									Y		Y	
Cottonseed meats/ dry:	B37Y	0.64	30	0.6														
Cottonseed meats/ rolled:	C47Y	0.56-0.64	30	0.6														
Cracklings/ crushed:	D47Y	0.64-0.8	30	1.3														
Cryolite/dust:	A38M	1.2-1.44	30A	2.0														
Cryolite/ lumpy:	D38	1.44-1.76	30A	2.1														
Cullet/ fine:	C39	1.29-1.92	15	2														
Cullet/ lump:	D39	1.28-1.92	15	2.5														
Dextrose:	A37VOW	0.64	N		N		N	-	-	-								
Diatomaceous earth/ natural:	A38XQJL	0.13	30A	1.6			-	-	Y	-					Y		Y	
Diatomaceous earth/ calcium:	A38XQJL	0.14					-	-	Y	-								
Diatomaceous earth/ flux:	A38XQJL	0.26																
Dicalcium phosphate:	A37	0.64-0.8	30	1.6														
/ granular:	C	0.96			30	17												
Disodium phosphate:	A37	0.4-0.5	30	0.5														
Distiller's grain/ spent/dry:	B37	0.4-0.48	30	0.5											Y		Y	
Distiller's grain/spent wet:	C47R	0.64-0.96	30	0.8														
Dolomite/ pulverized:	B38	1.04-1.28					-	-	Y	Y								
Dolomite/ crushed:	C38	1.28-1.6	30A	2														
Dolomite/lumpy:	D38	1.44-1.6	30A	2														
Earth/loam/ dry/ loose:	C38	1.22	30A	1.2	35	20												
Ebonite/ crushed:	C37	1.-1.12	30	0.8														
Egg powder	A37JLOW	0.26-0.34	30(N)	1.0	N		N	-	-	-								
Feeds/ soft: --															33	3.3	21	7
Feed/ bulk mash:	B36	0.56-0.64	30	0.6														
Feed/ molasses (5 to 8%):	B46	0.32-0.4	30	3														
Feed/ pellets and crumbles:	C46	1.58-0.64	30	0.6														
Feldspar, ground	A39	1.04-1.28	15	2			-	N	Y	Y					Y		N	
Feldspar/ lumps:	D39	1.44-1.6	15	2.0	32	17												
Feldspar powder:	A38	1.6	30A	2														
Feldspar screenings:	C39	1.2-1.28	15	2														
Ferric chloride:	HO	1.33	N		N		N	N	-	-					N		N	
Ferrous sulfate: green vitriol	C37O	0.8-1.2	30	1.0							Temp. Sensitive				Y		N	
Ferrous sulfide: (pyrite)	A38	1.68-1.9	30A	2.0														
Ferrous sulfide:	C28	1.9-2.16	30A	2														
Fish meal:	BWY	0.48-0.64													Y		Y	
Fish meal:	C47WY	0.56-0.64	30	1														
Fish scrap:	D47Y	0.64-0.8	30	1.5														
Flaxseed (Linseed):	B37Q	0.69-0.72	30	0.4	21	8									Y		Y	
Flaxseed/ rolled (linseed):		0.40			34	20												
Flaxseed cake expelled	D47	0.77-0.8	30	0.7														
Flaxseed meal:	B47	0.4-0.72	30	0.4	34	20									33	3.3	21	7
Flint:	B9	0.8-1.04					-	-	Y	Y					Y		N	
Flour/ wheat:	A47OMVW	0.53-0.64	30(N)	0.6	N		N	-	-	Y	1.5	180	4.5	65	27	5.3	11	12
Flue dust/ BOF:	A38MJ	0.72-0.96	30A	3.5														

continued

Table C-3 Solids: Particulate bulk density, characteristics and conveyability (continued).

Material	Size/code	Bulk Den Mg/ cu metre	Screw %	conve factor	Belt co repose o	tilt o	Bu el. ?	Fe ?	Bl tk ?	Air sld ?	Hydraul convey kg/kg	Pneumatic vert m/s	kg/kg	Conveying horiz m/s	kg/kg	vacuum m/s	kg/kg	lo press m/s	kg/kg
Flue dust/ blast furnace:	A38	1.76-2	30A	3.5															
Flue dust/ boiler house:	A37JM	0.48-0.72	30A	2															
Fluorspar: (calcium fluoride)																			
fluorspar/ micronized:	AT	0.93																	
(fluorspar)/ fine:	B38	1.28-1.44	30A	2.0			-	-	-	Y	_	3	525	9	130				
fluorspar/ lumps:	D38KT	1.44-1.76	30A	2															
Flyash:	A38J	0.48-0.72	30A	2.0			-	-	Y	Y									
flyash/ light	A37O	0.56																	
flyash/ sintered:	A39	0.69																	
flyash/ heavy:	9	1.09			N														
Fuller's earth/ dry raw:	A27	0.48-0.64	45	2.0	35	21						1.5	60	6	19	35	2.8	18	6.35
Fuller's Earth oily/ spent: (35 % oil)	C47N	0.95-1.05	30	2															
Fuller's Earth calcined:	A27	0.64	45	2															
Galena -100 M (lead sulfide)	A37T	3.85-4.2	30	--															
Gelatine/ granulated	B37OW	0.51	30	0.8															
Gilsonite	C37	0.59	30	1.5															
Glass/ batch	C39	1.28-1.6	15	2.5															
Glue/ vegetable/ powdered:	A47O	0.64	30	0.6															
Glue/ ground	B47O	0.64	30	1.7															
Glue/ pearl:	C37O	0.64	30	0.5															
Glue/ pellet:		0.72			25	11													
Gluten/ meal:	B37WO	0.64-0.69														33	3.3	21	7
Grains/ distillery/ dry/spent	D27L	0.48	30	0.4															
Granite/ fine	C29	1.28-1.44	15	2.5															
Grape pomace:	D47O	0.24-0.32	30	1.4															
Graphite/ flour:	A37JMW	0.45	30	0.5															
Graphite/ flake:	B27MW	0.64	15	0.5															
Graphite ore:	D37M	1.04-1.2	30	1															
Grass seed:	C28	0.16-0.5	30	0.4															
Gravel/ sharp:					40	27													
/round:					30	15													
Grits/ brewers:	B	0.67														Y		Y	
/ refined:	A															27	3.8	17	8.9
Guano/ dry:	C37	1.12	30	2															
Hay/ chopped:	C37LU	0.13-0.20	30	1.6															
Hexachlorocyclohexane: (Lindane)	A47T	0.9	30	0.6															
Hog fuel (bark) less than 10 cm																			
Hominy/ dry:	C27	0.56-0.80	45	0.4															
Hops/ spent/dry:	D37	0.56	30	1															
Hops/ spent/wet:	D47R	0.80-0.88	30	1.5															
Ice/ crushed:	D37N	0.56-0.72	30	0.4															
Ice /flaked:	C37N	0.64-0.72	30	0.6															
Ice/ cubes:	D37N	0.53-0.56	30	0.4															
Ice/ shell:	D47N	0.53-0.56	30	0.4															
Ilmenite ore:	A 9	1.76					-	-	-	Y									
Ilmenite ore:	D39	2.24-2.56	15	2															
Iron:	7Vu	3.77																	
Iron/ powder:	Vu	3.04																	
Iron ore concentrate	A39	1.92-2.88	15	2.2															
Iron ore/ soft:					35	21													
Iron oxide/ very light	7Q	0.3																	
Iron oxide/ light	7Q	0.64																	
Iron oxide/ medium	7QJ	1.28																	

Iron oxide/ heavy	7QJ	2.56																
Iron oxide pigment	A38JMW	0.40	30A	1.0	40	27												
Iron oxide/ limonite:		3.80			40	28												
Iron oxide/ millscale:	C38	1.2	30A	1.6														
Kaolin clay/talc mix	A37MJW	0.67-0.90	30	2														
Kaolin clay/ pulverized:	A	0.35			45	32												
Kaolin clay/ green crushed	D27	1.03	30	2.0	35	19												
Kieselguhr:	A	0.24									1.5	125	7.5	25				
Lactose	A37OW	0.50-0.70	30(N)	0.6	N		N	-	-	-								
Lead/ shot:	Tu	6.73	N		N		-	-	N	-	N		N		N		N	
Lead arsenate:	A37T	1.15	30	1.4														
Lead arsenite:	A37T	1.15	30	1.4														
Lead ore/ - 3. mm	B37T	3.2-4.3	30	1.4														
Lead ore/ - 12 mm	C38T	2.9-3.7	30A	1.4														
Lead oxide (red lead) -200M	A37MW	0.48-2.9	30	1.2														
Lead oxide -100M	A37W	0.48-2.4	30	1.2														
Lead oxide (red lead):	T	2.8-3.69			40	31												
Lead silicate/ granulated:		3.69			30	15												
Lead sulfate/ basic pulverized:		2.95			45	32												
Limonite ore/ brown:	C49	1.92	15	1.7														
Lime/unslaked / calcium oxide																		
Lime/ ground	B37OGwu	0.96-1.04	30	0.6	43	29	-	-	N	-	N		N		N		N	
Lime pebble:	C27OY	0.85-0.90	45(N)	2.0	30	17	-	-	Y	-	29	7			31	3.4	21	7.4
Lime/ spent dry carbide:	A	0.72			40	26												
Lime/ mason:		0.27			40	27												
Lime/ briquette:		0.96			26	15												
Lime, hydrated or slaked:																		
Lime/ hydrated pulverized:	A37JMGuOQ	0.35-0.64	30	0.6	Y		-	-	Y	Y								
Lime/ hydrated:	B37JM	0.48-0.64	30	0.8	42	30									27	3.9	12	14.8
Limestone/ Mineral calcium carbonate:																		
Limestone/ dust	A48JL	0.88-1.52	30A	2														
Limestone/ agricultural	B37	1.09	30	2														
Limestone/ crushed	D38	1.36-1.45	30A	2.0	47	34												
Limestone/ mixed sized:		1.68			35	21												
Limestone/ coarse/ sized:		1.57			25	12												
Lithopone:	A37JT	0.72-0.80	30	1														
Lucite	8	0.11																
Magnesium	9Vuw	1.07-1.60																
Magnesium carbonate/ light:	O	0.14			N													
magnesium carbonate	O	0.24																
magnesite/ dead/ burned	AK9	1.76	N								3	400	10	130				
Magnesium chloride	C47	0.53	30	1														
Magnesium oxide/ light	wuOQ	0.35																
Magnesium oxide/ medium	BwuOQ	0.72										Moisture affects						
Magnesium oxide/heavy	wuOQ	1.12																
Magnesium sulfate:	A37O	0.64-0.80	30	0.8														
Malt/ dry ground	B37VW	0.32-0.48	30	0.5											30	4.8	17	10.3
Malt/ meal	B27W	0.58-0.64	45	0.4														
Malt/ dry & whole	C37VOW	0.32-0.48	30(N)	0.5	N										Y		Y	
Malt/ sprouts	C37W	0.21-0.24	30	0.4														
Manganese	7VTu	1.52			39	24	-	-	N	-	N		N		N		N	
Manganese dioxide	A37VTG	1.12-1.36	30	1.5														
Manganese ore	D39	2.0-2.25	15	2														
Manganese oxide	A38	1.92	30A	2														
Manganese sulfate	C39(8)	1.12	15	2.4														
Marble/ crushed	B39(8)	1.28-1.52	15	2														
Margarine	E47YPWQ	0.95	30	0.4														
Marl (clay)	D38	1.28	30A	1.6														
Meat/ ground	E47YXGQ	0.80-0.88	30	1.5														
Meat/ scrap without bone	E48Y	0.64	30A	1.5							cleanout often				Y		Y	

continued

Table C-3 Solids: Particulate bulk density, characteristics and conveyability (continued).

Material	Size/code	Bulk Den Mg/ cu metre	Screw %	conve factor	Belt co repose o	tilt o	Bu el. ?	Fe ?	Bl tk ?	Air sld ?	Hydraul convey kg/kg	Pneumatic vert m/s	kg/kg	Conveying horiz m/s	kg/kg	vacuum m/s	kg/kg	lo press m/s	kg/kg
Metals/ light/ granular	9Vu	1.12					-	-	N	-		N		N		N		N	
Metals/ heavy/ granular		3																	
Metal ores	9V	2.7	N				-	-	N	-		N		N		N		N	
Mica/ pulverized	A38J	0.21-0.24	30A	1															
Mica/ very thin	SX	0.19	N				N	-	N	-		N		N		N		N	
Mica/ flakes	B18JL	0.27-0.35	30A	1															
Mica/ ground	B38	0.21-0.24	30A	0.9	36	23													
Mica ore	SX	0.53	N				N	-	N	-		N		N		N		N	
Milk/ malted	A47WQ	0.43-0.48	30	0.9															
Milk/ sugar	A37QW	0.51	30	0.6															
Milk/ dried flake	B37WOL	0.08-0.10	30	0.4															
Milk/ powdered	B27	0.32-0.72	45	0.5												Y		Y	
Milk/ whole & powdered	B37WOQV	0.32-0.58	30(N)	0.5	N		N	-	-	-									
Mill scale (steel)	E48G	1.92-2.0	30A	3															
Milo/ ground	B27	0.51-0.58	45	0.5															
Milo (maize, Kafir)	B17V	0.64-0.72	45	0.4															
Mineral ores/ light	9KVu	1.12	N				-	-	N	-		N		N		N		N	
Mineral ores/ heavy	9KVu	2.25	N									N		N		N		N	
M & M candy		0.8																	
Molybdenum	7VJ	2.25																	
Molybdenum disulfide		1.27																	
Molybdenite powder	B28	1.71	30A	1.5	40	25													
Monosodium phosphate	B38	0.80	30A	0.6															
Mortar/ wet	E48G	2.40	30A	3															
Mustard seed	B17V	0.72	45	0.4															
Naphthalene flakes	B37	0.72	30	0.7															
Niacin (Nicotinic acid)	A37W	0.56	30	0.8															
Nickel	9	8.6																	
Oats/ flour	A37	0.56	30	0.5								Sluggish				Y		Y	
Oats/ crushed	B47VL	0.35	30	0.6															
Oats/flou	B	0.32										Sluggish				Y		Y	
Oats/ hulls	B37VL	0.13-0.19	30	0.5															
Oats	C27JV	0.40	45	0.4	21	8										30	3.9	17	8.3
Oats/ crimped	C37	0.30-0.42	30	0.5															
Oats/ rolled	C37VL	0.30-0.38																	
Oats/ ground	VOXW	0.43	N		N		N	-	N	-		N		N		N		N	
Onions/ dehydrated		0.35																	
Orange peel/ dry	E47	0.24	30	1.5															
Organic materials	5XW	0.64	N		N		N	-	N	-		N		N		N		N	
Oxalic acid crystals (ethane diacid)	B37XH	0.96	30	1															
Oyster shells/ ground	C38G	0.80-0.96	30A	1.8															
Oyster shells/ whole	D38GR	1.28	30A	2.3															
Paper pulp/ < 4%	E47	1.0	30	1.5															
Paper pulp/ 6 to 15%	E47	0.96-1.0	30	1.5															
Paraffin cake/ -12 mm	C47P	0.72	30	0.6															
Peanut meal	B37W	0.48	30	0.6															
Peanuts/ shelled	C37X	0.56-0.72	30	0.4			-	-	N	-		N		N		N		N	
Peanuts/ unshell/unclean	D38X	0.24-0.32	30A	0.7								Control velocity				Y		Y	
Peanuts/clean in shell	D37X	0.24-0.32	30	0.6															
Peas/ dried	C17X	0.72-0.80	45	0.5															
Perlite/ fine bead		0.13-0.24										Control velocity				Y		Y	
Perlite/ expanded	C38	0.06-0.11	30A	0.6														Use fan	
Phenolic resin	VJ	0.48				18	N	-	N	-		18		N		N		N	

Phosphate acid fertilizer	B27G	0.96	45	1.4														
Phosphate rock/ pulverized	B38	0.96	30A	1.7	40	28												
Phosphate rock/ ground	B 7	1.28-1.60			27	14	-	-	Y	Y								
Phosphate rock/ broken	D38KO	1.2-1.36	30A(N	2.1			-	-	N	-	N		N		N		N	
Phosphate rock/pulverized	A	1.28									3	260	10	90				
Phosphate sand	B39	1.44-1.60	15	2														
Phthalic anhydride/flakes/		0.67			24	10												
Pigments	5WQ	0.54	N				N	-	N	-	N		N		N		N	
Pitch/granular	7VUOX	0.65	N		27	13	N	-	N	-	N		N		N		N	
Pitch/ powder	7GUOJ	0.34					N	-	-	-								
Plastic powder	SVJ	0.45					N	-	N	-	N		N		N		N	
Plastic pellets	SVXz	0.56																
Polyethylene/ beads	C47X	0.48-0.56	30	0.4											24	7.4	21	12
Polyolefin pellets		0.53	21	29	10						21	10			29	10		
Polystyrene beads	B37WX	0.64	30	0.4														
Polyvinyl chloride/ powder	A47PG	0.32-0.48	30	1.0	27						27	0.6		0.8				
Polyvinyl chloride/ beads	E47PWXG	0.32-0.48	30	0.6	20													
Porcelain	7	1.2																
Potassium carbonate	B38	0.82	30A	1														
Potassium chloride:																		
inpure/ dry	B39	1.12	15	2														
mine run	D39	1.20	15	2.2														
Potassium chloride pellets	C27GO	1.92-2.1	45	1.6														
Potassium nitrate - 3 mm	B28GV	1.28	30A	1.2														
Potassium nitrate -12 mm	C18GV	1.22	30A	1.2														
Potassium sulfate	B48Q	0.67-0.77	30A	1														
Potato flour	A37JVW	0.77	30	0.5														
Potato flakes		0.13-0.21																
Prunes/ whole		0.61																
Pumice -3 mm	B48	0.67-0.77	30A	1.6														
Pyrite/ pellets	C28Vu	1.92-2.08	30A	2.0							N		N		N		N	
Quartz -100 M	A29	1.12-1.28	15	1.7														
Quartz -12 mm	C29	1.28-1.44	15	2														
Raisins/ dry	2W	0.5																
Rice/ bran	B37LV	0.32	30	0.4														
Rice/ grits	B37W	0.67-0.72	30	0.4														
Rice/ hulls	B37LV	0.32-0.34	30	0.4														
Rice/ polished	C17W	0.48	45	0.4														
Rice/ hulled	C27W	0.72-0.79	45	0.4	20	8												
Rice/ rough	C37V	0.51-0.58	30	0.6							lo vel. + hi solids				Y		Y	
Rice Krispies		0.16																
Rosin/ - 12 mm	C47X	1.04-1.09	30	1.5														
Rubber/ reclaimed ground	C47	0.37-0.80	30	0.8	35	22									33	3.6	25	--
Rubber/ pellets	D47	0.80-0.88	30	1.5														
Rye	B17V	0.67-0.77	45	0.4	32	18					25				31	4.8	-	-
Rye bran	B37L	0.24-0.32	30	0.4														
Rye feed	B37V	0.53	30	0.5														
Rye meal	B37	0.56-0.64	30	0.5														
Rye middlings	B37	0.67	30	0.5														
Rye shorts	C37	0.51-0.53	30	0.5														
Safflower cake	D28	0.80	30A	0.6														
Safflower meal	B37	0.80	30	0.6														
Safflower seed	B17V	0.72	45	0.4														
Salicylic acid	B39O	0.46	15	0.6														
"Salt"/ dry		1.4													36	2.6	-	-
Salts/ acid	5GHO	0.72	N				N	-	-	-								
salts/ alkaline/ powder	5GOJ	0.77																
salts/ alkaline/ granular	GO	0.88																
Salt/ rock: dry/ coarse	C38GO	0.72-0.96	30A(N	1.0	25	11	N	-	-	-								
rock/ kiln dried	COH	0.68-0.80									3	200	9	65	36	2.6	--	3.3
	KXH	1.12	N		N													

continued

Table C-3 Solids: Particulate bulk density, characteristics and conveyability (continued).

Material	Size/code	Bulk Den Mg/ cu metre	Screw %	conve factor	Belt co repose o	tilt o	Bu el. ?	Fe ?	Bl tk ?	Air sld ?	Hydraul convey kg/kg	Pneumatic vert m/s	kg/kg	Conveying horiz m/s	kg/kg	vacuum m/s	kg/kg	lo press m/s	kg/kg
Salt/ table	HOW	0.96	N		N		N	-	-	-									
Salt/ damp	HOW	0.87	N		N		N	-	N	-		N		N		N		N	
Salt/ damp and wet	9	1.57	N		N		N	-	N	-		N		N		N		N	
Salt/ dry & fine	B38GOH	1.12-1.28	30A	1.7	31	16										Y		Y	
Sand/ resin coated zircon	A29	1.84	15	2.3															
Sand/ dry bank (damp)	B49	1.76-2.1	15	2.8	45	33	-	-	N	-		N		N		N		N	
Sand dry bank (dry)	B39	1.44-1.76	15	1.7	32	18						22	6						
Sand/ dry silica	B29	1.44-1.6	15	2															
Sand/ resin coated silica	B29	1.67	15	2															
Sand/ foundry (shake out)	D39	1.44-1.6	15	2.0	30	16													
Sawdust/ dry	B47OQ	0.16-0.21	30	0.7															
Sea coal	B38	1.04	30A	1															
Sesame seeds	B28	0.43-0.66	30A	0.6															
Shale/ crushed	C38	1.36-1.44	30A	2.0	39	26													
Shellac/ powdered/ granulated	B37W	0.50	30	0.6															
Silica/ flour	A48	1.2-1.28	30A	1.5			-	-	Y	Y									
Silica/flour	A	0.8-0.96										1.5	260	6	65				
Silica/ very light	9J	0.08			N		N	-	-	-									
Silica/ light	9J	0.24			N														
Silica gel 12 to 75 mm	D39YPXO	0.72	15	2															
Slag/ furnace/ granular/ dry	C39	0.96-1.05	15	2.2															
/ blast furnace/ crushed	D39L	2.08-2.9	15	2.5	25	13													
Slate/ ground - 3 mm	B38	1.3-1.36	30A	1.6	35	22													
Slate/ crushed -12mm	C38	1.28-1.44	30A	2															
Sludge/ sewage/ dried/ground	B48H	0.72-0.88	30A	0.8															
Sludge/ sewage/ dried	E49G	0.64-0.8	15	0.8															
Snow/ fresh		0.08-0.2	30	0.4															
/packed		0.24-0.56	30	0.8															
Soap/ beads or granules	B37X	0.24-0.56	30	0.6															
Soap/ chips	C37X	0.24-0.4	30	0.6	30	18	Y	-	N	?				N		N		N	
Soap/ flakes	B37XQL	0.08-0.24	30	0.6															
Soap/ powder	B27Q	0.32-0.4	45	0.9															
Soap/ detergent	B37KX	0.24-0.8	30	0.8															
Sodium aluminate/ ground	B38	1.15	30A	1															
Sodium aluminum sulfate	A38	1.20	30A	1															
Sodium bicarbonate: (Baking soda)	A27	0.64-0.88	45	0.6															
Sodium carbonate:																			
light	A38LGOJu	0.32-0.56	30A	0.8								3	195	9	65	33	3.45	20	7
dense/	B38G	0.88-1.04	30A	1.0								3	120	12	38	N		N	
naturally mined	B	0.79					-	-	Y	-						Y		Y	
Sodium hydroxide:	B37HTO	0.96-1.12	30	1.8			-	-	Y	-									
/flakes	C47HLTO	0.75	30	1.5															
Sodium nitrate:																			
fine		1.4			31	18													
granular	D27HVOu	1.12-1.28	30(N)	1.2	24 (N)	10	N	N	N	-		N		N		N		N	
Sodium perborate	B	0.87										3		10	38				
Sodium phosphate	A37	0.8-0.96	30	0.9															
Sodium silicate	5HO	0.71																	
Sodium sulfate (salt cake)																			
dry/ coarse	B38OG	1.36	30A	2.1	31	18													
dry/ pulverized	B39O	1.04-1.36	30A	1.7								3		10	65	Y		Y	
pure (Glauber's salts)	B	1.41														Y		Y	
Sodium sulfite	B48QO	1.2-1.55	30A	1.5												Y		Y	
Sodium tripolyphosphate	O	0.96																	

Soybean/ flour	A37JV	0.43-0.48	30	0.8													
Soybean/ meal/cold	B37O	0.64	30	0.5						max oil 13% %				33	3.3	21	7
Soybean meal/ hot	B37G	0.64	30	0.5													
Soybeans/ whole	C28V	0.72-0.8	30A	1.0										Y		Y	
Soybean/ cracked	C38V	0.48-0.64	30A	0.5													
Soybean/ raw flakes	C37L	0.29-0.4	30	0.8													
Soybean/ cake	D37	0.64-0.7	30	1													
Spices	VOW	0.56	N			N	-	-	-	N		N		N		N	
Starch	A17JV	0.4-0.8	45	1.0										27	3.8	17	8.9
pellets <6mm	B V	0.63-0.8															
Starch, corn	VOJ	0.56	N			N	-	-	-								
Starch, corn, pearl	VOJ	0.72	N			N	-	-	-								
Starch/ potato	VOJ	0.72	N			N	-	-	-	N		N		N		N	
Starch/ tapioca		0.63															
Stearine/ beads	C	0.54										15	7.5				<7.5
Steel turnings/ crushed	D48R	1.6-2.4	30A	3													
Sugar beet pulp/ dry	C28	0.19-0.24	30A	0.9													
Sugar beet pulp/ wet	C37Q	0.4-0.72	30	1.2													
Sugar/ refined/																	
granulated/dry	B37WO	0.8-0.88	30	1.1		-	-	Y	Y					33	3.3	18	7.8
granulated & wet	C37Q	0.88-1.04	30	1.7													
powdered	A37QW	0.8-0.96	30	0.8													
raw	B37QW	0.88-1.04	30	1.5		-	-	N	-	N		N		N		N	
sugar XXX	5VOW	0.48	N			N	-	N	-	N		N		N		N	
Sulfur/ powdered	A37JV	0.8-0.96	30	0.6													
crushed - 12mm	C37V	0.8-0.95	30	0.8													
lumpy -75 mm	D37GQ	1.28-1.36	30(N)	0.8	N	N	-	-	-	N		N		N		N	
Sunflower seed	C17	0.3-0.6	45	0.5													
Super phosphate	KO	0.96	N														
Talc: (Magnesium silicate/ hydrous)																	
light	OJ	0.19			N												
medium	OJ	0.45			N												
heavy	OJ	0.72			N												
powder	A47LQ	0.74-0.99	30	2.0		-	-	-	Y	buildup in pipes				Y		Y	
Talcum -12mm	C38	1.28-1.44	30A	0.9													
/ powder	B38J	0.8-0.96	30A	0.8													
Tanbark/ ground	B47	0.88	30	0.7													
Timothy seed	B37LV	0.58	30	0.6													
Tin	8	3.74				N	-	N	-	N		N		N		N	
Titanium	9	1.6															
Titanium sponge	9KV	0.72															
Titanium dioxide/ light	7QJ	0.50								moisture sensitive				?		?	
Tobacco snuff	B47	0.48	30	0.9													
/scraps	D47	0.24-0.4	30	0.8													
Tricalcium phosphate	A47	0.64-0.8	30	1.6													
Triple super phosphate	B38HT	0.8-0.88	30A	2													
Tripolyphosphate:	B	1.05								1.5	390	7.5	65				
Trisodium phosphate (oakite)																	
pulverized	A38	0.80	30A	1.7		-	-	Y	Y	1.5		7.7	65	33	3.2	23	7
granular	B38	0.8-1.04	30A	1.7		-	-	Y	Y					33	3.2	23	7
	C38	0.96	30A	1.7													
Tung nut/ meats crushed	D27	0.45	30	0.8													
whole	D17	0.4-0.48	30	0.7													
Tungsten	9V	5.13				N	-	N	-	N		N		N		N	
Uranium dioxide	A	3.5								7		18	131				
Urea/ prilled & coated	B27u	0.69-0.74	45	1.2													
Vermiculite/ expanded	C37L	0.26	30	0.5													
ore	D38	1.28	30A	1													
Vetch	B18V	0.77	30A	0.4													

continued

Table C-3 Solids: Particulate bulk density, characteristics and conveyability (continued).

Material	Size/code	Bulk Den Mg/ cu metre	Screw %	conve factor	Belt co repose o	tilt o	Bu el. ?	Fe ?	Bl tk ?	Air sld ?	Hydraul convey kg/kg	Pneumatic vert m/s	kg/kg	Conveying horiz m/s	kg/kg	vacuum m/s	kg/kg	lo press m/s	kg/kg
Walnut shells/ crushed	B38	0.56-0.72	30A	1															
Wheat/ germ	B27	0.29-0.45	45	0.4								25	10			31	4.8	--	--
cracked	B27VOW	0.64-0.72	45	0.4															
grain	C27V	0.72-0.77	45	0.4								25	10			31	4.8	--	--
Whey	5W	0.56	N		N		N	-	N	-		N		N		N		N	
White lead/ dry	A38JT	1.2-1.6	30A	1															
Wood chips/ screened	D47LR	0.16-0.48	30	0.6	35	25										N		Y	
flour	B37GVUOQ	0.26-0.58	30	0.4			N	-	-	-						30	3.3	16	10
shavings	E47LR	0.13-0.26	30	1.5															
Zinc/ concentrate residue	B39	1.2-1.25	15	1															
Zinc	Vu	3.08		N			N	-	N	-		N		N		N		N	
Zinc/sintered ore	9	1.84			38	27													
Zinc oxide/ light	A47LQ	0.16-0.24	30	1															
light	A47QJ	0.48-0.56	30	1.0			-	-	-	Y						Y		Y	
sintered	7	1.6																	
Zinc stearate	5QJ	0.13																	
Zircon/	9y	2.64																	
milled	9	1.82																	

Table C-4 Solids: density, hardness, work index, zpc, electrostatic and magnetic properties

Type of solid or mineral	formula	Mass density Mg/cu metre	Hardness Moh scale	Work index $kWh(\mu m)^{0.5}$/Mg	zpc pH	rel permitt	Thresh volts kV/cm	S	Magnetism relative to iron = 100	Rationalized susceptabilities volume/ cu. metre	mass/ cu. metre
Actinolite	CaFe3Mg3O12Si4	3.0-3.2	5.0-6.0			6.82	5.4	E			
Adularia									-0.0004		
Albite	AlO8NaSi3	2.6-2.7	6.0-6.6		2						
Albumin					4.8						
Algae					2.6						
Alumina		3.9		19.3							
Aluminum											
Aluminous oxide							6.1	P	D	-0.16×10^{-6}	
Aluminum oxide											
a/synthetic					9						
					8						
amorphous					8						
mineral					9						
Alunite	Al12H6K2O22S4	2.7	3.8								
Amphiboles		2.9-3.4	5.5-6.5				3.2	N			
Anatase	O2Ti										
/synthetic											
Andesite		2.84		24.4							
Anglesite	PbSO4	6.1-6.4	2.8-3.0								
Anhydrite						6.09	3.5	P	NM		
Anthophylite	(FeMg)SiO3	3.0-3.2	5.0		4						
Anthracite					8	33 to 81	1.6	E			
Antigorite					ND						
Antimonite	Sb2S3	4.6-4.7	2								
Antimony	Sb						3.5	E	-0.0023		
Antimony oxide	O5Sb2				<1						
Apatite	Ca5(F/Cl)P3O12	3.2	5		8.5	5.72	5.3	P	0.083 to -0.0034		-32×10^{-9}
Aragonite	CaCO3	2.9	3.5-4.0		<5	7.44	6.7	P			
Argentite	Ag2S	7.2-7.5	2.-2.5			> 81			0.102		
Argonite									-0.0048		
Arsenopyrite	AsFeS	5.9-6.2	5.5-6.0			> 81			0.054		
Arsenic						10.23	3.0	E			
Augite					3				0.027		
Axinite						6.15	4.6	N			
Azurite	Cu3C2H2O8	3.8-3.9	3.5-4.								
Barite	BaO4S	4.3-4.7	2.5-3.5	6.86	3.4	7.86	2.6	E	NM		
/synthetic											
Basalt/ altered		2.9		22.45						altered	
Basinasite						NC			WM		
Bauxite	Al2O6H6	2.38	1-3	9.66/10.4		10.85	3.9	N			
Bayerite					9.3						
Bentonite/clay	(CaMg)(AlFe)2O6Si	2.1	1		<3		1.6	E			
Beryl	Al2Be3O18Si6	2.6-2.8	7.5-8		3.5	NC			0.0008		
Beryllium oxide					10.2						
Biotite/ mica	(HK)2(MgFe)2Al2O12Si3	2.7-3	2.5-3		1	9.28	2.2	E	0.115 to 0.0032		
Bismuth	Bi	4.4				> 81	2.1	E	-0.0032		
Boehmite	-AlHO2	3-3.1	3.5		<2						
/synthetic					8						
Borax	B4Na2H20O17	1.7	2-2.5								
Bornite	Cu5FeS4	4.9-5.4	3-3.5			> 81	2.1	E	0.086		
Braunite	Mn7O12Si	4.7-5	6-6.5			> 81					
Brookite						C			NM		

continued

Table C-4 Solids: density, hardness, work index, zpc, electrostatic and magnetic properties

Type of solid or mineral	formula	Mass density Mg/cu metre	Hardness Moh scale	Work index $kWh(\mu m)^{0.5}$ /Mg	zpc pH	rel permitt	Thresh volts kV/cm	S	Magnetism relative to iron = 100	Rationalized susceptabilities volume/ cu. metre	mass/ cu. metre
Bronzite					3.3						
Brucite	H2MgO2	2.4	2.5								
/synthetic		2.4			12						
Cadmium hydroxide	CdH2O2	4.8			>10						
Cadmium oxide	CdO	6-8			10.4						
Calamine	H2O5SiZn2	3.4-3.5	4.5-5						0.187		
Calcite	CaCO3	2.6-2.8	2-3		8.5	6.36	4.9	P	0.013 to -0.0004	-13×10^{-6}	-4.8 to -5.2×10^{-9}
/synthetic		2.7			9.5						
Calcium phosphate (tertiary)	Ca2O8P2	3.14			7.6						
Carnotite	K2O12U4V2		1.5								
Cassiterite	O2Sn	6.8-7.1	6-7		4-7	27.75(C)			NM		3.7 to 8.8×10^{-9}
/synthetic	O2Sn	7.0			4.5						
Celestite	O4SrS	3.9-4	3-3.5		2.3	6.94			0.038 to -0.0005		
Cement/ clinker		3.15		14.84							
/ raw material for		2.67		11.63							
Cerargyrite									0.105		
Cerium dioxide					6.7						
Cerussite	CO3Pb	6.4-6.6	3-3.5			5.47					
Chalcocite	Cu2S	5.5-5.8	2.5-3			> 81	3.0	E	0.038		
Chalcopyrite	CuFeS2	4.1-4.3	3.5-4			> 81	2.1	E	0.051 to NM	314×10^{-6}	10 to 70 E-9
Chert	O2Si	2.6	7				4.0	N			
Chlorite					4-5						
Chromite	Cr2FeO4	4.0-4.8	5.5-7	10.56	7	11.03(C)	2.5	E	WM		
Chromium oxide	Cr2O3				7				SM	840 E-6	
Chrysocolla	CuH4O5Si	2.-2.2	2-4		2						
Chrysolite	(MgFe)2O4Si	3.3	6.5-7		>12	approx 35	4.1	P			
Cinnabar	HgS	8-8.2	2-2.5			8.43			0.038		
Clay		2.23		7.8/6.9		8					
/bentonite							1.6	E			
/calcined		2.32		1.57							
/ferrogenous											350 to 560 E-9
/kaolinite					4-7	11.18	3.0	N			
Coal					1-2						
/anthracite					8		1.6	E			
/bituminous		1.3		14.3/12.54			1.8	P			
Coal/ river (anthracite)											
Coal/ culm (anthracite)											
Cobalt hydroxide	CoH2O2				11						
Cobaltite	AsCoS	6-6.3	5.5						0.0023		
Coke	C	1.51		22.77							
Coke/ petroleum		1.78		81.2							
Columbite	(FeMn)(NbTa)2O6	6.3	6			C			WM		
Copper hydroxide	CuH2O2	3.4									
Copper ore		2.95		12.9/14.4							
Copper/ native	Cu	8.8	2.8			> 81			NM		
Copper oxide	CuO	6.4							SM	240 E-6	
Coral		2.7		11.2							
Corundum	Al2O3	3.9-4.1	9			5.35	6.2	E	0.264 to -0.0006/NM		
Covellite	CuS	4.7	1.5-2			> 81					
Cristobalite											
Cryolite	AlF6Na3	3	2.5			8.13	3.8	P	0.019		
Cummingtonite											
Cuprite	Cu2O	5.9-6.1	3-3.5			16.20			0.51 to 0.0096		

Davidite						C			M		
Dental enamel					4						
Descloizite	H2O10RV2	6	3.5								
Diabase/ altered											
Diabase/ fresh											
Diamond	b-C	3.5	10			4.58 ?(C)			NM		
Diaspore	a-AlHO2	3.3-3.5	6-7		<2						
/synthetic					6						
Dioptase									0.012		
Diorite		2.7		21.34							
Diorite/ augite											
Disthene	Al2O5Si	3.2-3.7	7-7.5								
Dolomite	(Ca/Mg)CO3	2.8-2.9	3.5-4	12.43	<7	8.45	3.7	P	0.178 to 0.057/0.001		
Eicosanoic acid					3						
Eggonite	AlH4O6P				4						
Emery		3.48		64.							
Enargite	AsCu3S4	4.4-4.5	3-3.5			> 81			0.019		
Enstatite						8.23	3.5	N			
Enzymes											
2-a-N acetylhexosamidases					5.5						
ATP from calf liver					10.6						
ATP from rabbit muscle					9.6						
n-acetyl galactosaminidase					4.5						
a-galactosidase					4-5						
a-N-acetylglucosaminidase					3-6						
isomerase					6.7						
Epidote	Ca2(AlOH)(AlFe)2O12Si3	3.2-3.5	6-7			NC			WM		
Euxenite						C			WM		
Fayalite (fayerite)	Fe2O4Si				4-5					5.4 E-3	1.26 E-6
Feldspar		2.59		12.84		NC			NM	20 E-6	
/microcline					2-4	6.92	3.4	E			
/see orthoclase									0.0023		
/see plagioclase											
Ferberite	FeO4W	7.2-7.5	5-5.5						0.101	502 E-6	
Ferric hydroxide											
/amorphous	FeH3O3				8.5						
Ferric oxide/a-mineral	Fe2O3	5.24			4-7						
/synthetic					8.6						
-synthetic					6.7						
Ferro-chrome		6.75		9.79							
Ferro-manganese		5.9		8.58							
Ferrosilicon		4.9		14.1							
Flint		2.65		28.82							
Fluorapatite					7						
/synthetic					6-7						
Fluorite	CaF2	3-3.2	4		6.2	7.11	2.3	E	0.054/NM to -0.0004		
Fluorspar		3		10.78							
Forsterite					3-9						
Franolite					3-5						
Franklinite	(ZnFeMn)(FeMn)2O4	5.2	5.5-6.5			9.37	3.7	E	13.09 to 1.48	31 to 46 E-3	
Gabbro		2.83		20.35							
Gadolinium oxide									WM	13 E-3	
Galena	PbS	7.4-7.6	2.5	11.2		> 81	3.7	E	0.019 to -0.0011		-4.4 E-6 to -4 E-9
Garnet		3.3	6.5-7.5	13.64	4.4	NC	8.1	E	0.149	5 E-3	
Geothite	a-FeHO2	4.0 to 4.4	5-5.5		3-7	11.7				1.4 E-3	325 E-9
/synthetic					4-7						
Gibbsite	AlH3O3	2.35	2.5-3.5		5	8.37					
Glauconite					2						
Glass		2.58		3.41							

continued

Table C-4 Solids: density, hardness, work index, zpc, electrostatic and magnetic properties

Type of solid or mineral	formula	Mass density Mg/cu metre	Hardness Moh scale	Work index $kWh(\mu m)^{0.5}$ /Mg	zpc pH	rel permitt	Thresh volts kV/cm	S	Magnetism relative to iron = 100	Rationalized susceptabilities volume/ cu. metre	mass/ cu. metre
Gneiss/ biotite		2.7		22.11							
/ granite											
/ hornblende											
Gold/ native	Au	16-19	2.5-3			> 81			NM		
/ pure	Au	19.3									
/ ore		2.86		16.28							
Granite		2.72		15.8/18.3							
/ hornblende											
/ biotite		0.8 to 1.0									
Graphite	a-C	2.1-2.2	1	49.5	2-4	> 81	1.3	E	-0.032 to -0.0056		
/plumbago							1.6	E			
Gravel		2.68		27.5							
Gypsum	CaH4SO6	2.3	1.5-2		2-10	6.83	3.4	P	0.038 to 0.016/ NM		
/rock		2.69		9.0/ 7.4							
Hair	human				3	human					
Halite	NaCl	2.1-2.2	2			6.3 to 7.3	1.8	E	-0.0004		
Hematite	a-Fe2O3	4.9-5.3	5.5-6.9	9.4 to 14.1		> 81	2.8	E	0.77/0.26	1.4 to 10 E-3	260 E-9 to 2.3 E-6
Hemimorphite							4.1	E			
Hornblende		3.2	5-6		ND	NC			0.025/WM		
Huebnerite	MnO4W	7.2-7.5	5-5.5						0.105		
Hydroxyapatite					7						
/synthetic					8.5						
Ilmenite	FeTiO3	4.7	5-6	14.4		34 to 81	3.2	E	9.15	7 to 18 E-3	1.5 to 3.8 E-6
Iron									100		
Iron carbonate									WM	4.7 E-3	
Kaolinite/clay	Al2H4O9Si2	2.6	2-2.5		4-7	11.18	3.0	N			
Kieserite	H2MgO5S	2.6	3.3			8.2					
Kisselglas/synthetic					2.5						
Kyanite	Al2O5Si	3.2-3.6	4-7	20.8		7.18	4.1	E	NM		
Labradorite						6.98	2.3	E			
Lanthanum oxide					10.4						
Lead hydroxide	H2O2Pb	7.6			9-11						
Lead ore		3.44		12.54							
Lead/zinc ore		3.37		12.49							
Lepidocrocite	FeHO2				7.4						
/synthetic					5-7						
Lepidolite	Al2F2H2KLiO11Si3	2.8-3.3	3		1.8	7.36	2.3	E			
Limestone		2.7		12.8/14.0						7.5 E-5	
Limonite	FeHO2 or Fe4H6O9	3.3-4	5-5.5			6.95	3.9	E	0.32 to NM	9 E-3	0.3 to 0.5 E-6
Lollingite	AsFeS2	7.4-7.5	5-5.5								
Magnesium hydroxide, see Brucite											
Magnesium oxide	MgO	3.65			12.4						
Magnesite	CMgO3	2.9-3.1	4-4.5	18.5		6.99	3.9	P	0.019		
/synthetic	CMgO3	3.04			6-6.5						
Magnetite	Fe3O4	4.9-5.2	5.5-6	11.2/14.3	6.5	34 to 81	3.5	E	48 to 15	1.5 to 38	0.3 to 7.6 E-3
/synthetic		5.2			6.5						
Malachite	CCu2H2O5	4	3.5-4			6.23					
Manganese dioxide	MnO2										
/synthetic		5.03			4-4.5						
/ -					7.2						
/d-					2.8						
Manganese hydroxide	H2MnO2	3.26			7						
Manganese oxide	Mn2O3	4.8							WM	5 E-3	1 E-6

Manganese nodules											0.9 to 3.3 E-6
Manganese ore		3.74		13.75							
Manganite	g-MnHO2	4.2-4.4	4			> 81	5616	E	0.194	6 E-3	
Marcasite	FeS2	4.8-4.9	6-6.5			34 to 81	5460	E			
Marlite											
Mercuric oxide	HgO	11.14			7.3						
Metallics/ powdered iron											
Mica		2.89		147					5.9 to 0.0032		
see biotite											
see muscovite											
Microcline, see Feldspar											
Millerite	NiS	5.2-5.6	3 to 4								
Molybdenite	MoS2	4.7-5	1-1.5		1	> 81	3.2	E	0.118 to 0.05/ NM		
Molybdenum		2.7		14.3							
Monazite	CePO4	4.9-5.5	5-5.5			7.98	3.0	E			
Montmorillonite											
Muscovite/ mica	Al3H2KO12Si3	2.8-3	2-2.5		4	10.00	1.3	P	NM		
Naphthalene									D	-10 E-6	
Nephelite	AlNaO4Si	2.5-2.7	5.5-6			6.82	2.8	E			
Niccolite	AsNi	7.3-7.7	5-5.5			approx 33	3.5	E	0.016		
Nickel hydroxide	H3NiO3				12						
Nickel oxide	NiO	7.45			10.3						
Nickel ore		3.32		13.1							
Oil					2						
Oil shale		1.76		19.9							
Oligoclase						6.37	2.8	N			
Olivine	(MgFe)2O4Si	3.3	6.5-7		3-11				WM		
Orpiment									0.089		
Orthoclase	AlKO8Si3	2.5-2.6	6		4.8	6.20			0.035 to 0.0032		
Pentlandite	(Fe/Ni)S	4.5-5	3.5-4.5								
Peridotite											
Perlite											
Perovskite						NC			NM		
Petzite	(AuAg)2Te	9.1	2.5								
Phosphate rock	Ca3O8P2	2.6-3.2	5	11.14							
Plagioclase	Al(Na/Cl)O8Si3	2.6-2.8	6-6.5								
Platinum oxide					>14						
Plutonium dioxide					9						
Polystyrene latex											
Polytetrfluoroethylene					3						
Porphyries											
Potash ore		2.37		9.8							
Psilomelane	MnO2	4.4-4.7	4-6								
Pumice		1.96		13.12							
Pyrite	FeS2	4.9-5.2	6-6.5		6.5	34 to 81	3.5	E	0.203 to 0.002	0.2 to 2 E-3	0.04 to 0.4 E-6
/concentrate											
/ore		3.48		9.8							
Pyrochlore											
Pyrophyllite	AlHO6Si3	2.8-2.9	1-2		1.7						
Pyrolusite	MnO2	4.7-5	3.5-4.5		5.6	> 81	2.1	E	0.028 to 0.025	8 E-3	
Pyroxene	Ca(AlFeMnMg)O6Si3	3.3	5-6		2.8	2.7	N				
Pyrrhotite	FeS	4.6-4.7	4		2	> 81	3.0	E	2.5	4 to 72 E-3	
/ore		4.0		10.56							
Quartz	SiO2	2.5-2.8	7	14.05	1.9	6.53	4 to 6.3	N	0.175 to -0.0005/NM	-16 E-6	5.4 E-9 to -6.2 E-9
/ore				14.93							
Quartzite/ pyroxene		2.7		13.4/19.1							
Rhodochrosite	CMnO3	3.3-3.7	3.5-4.5			6.77	3.9	E			
Rhodolite						7.4	P				
Rhodonite	MnSiO3	3.4-3.7	5.5-6.5		2.8	7.10			0.56		

continued

Table C-4 Solids: density, hardness, work index, zpc, electrostatic and magnetic properties

Type of solid or mineral	formula	Mass density Mg/cu metre	Hardness Moh scale	Work index $kWh(\mu m)^{0.5}$ /Mg	zpc pH	rel permitt	Thresh volts kV/cm	S	Magnetism relative to iron = 100	Rationalized susceptabilities volume/ cu. metre	mass/ cu. metre
Rhyolite											
Rutile	TiO2	4.2-4.3	6-6.5		5-6.5	(5.8?) VC	7332	E	0.168 to 0.0034		
/ore		2.84		13.42							
Salt/ rock see Halite											
Samarskite						C			WM		
Sandstone		2.68		12.68							
/ calcareous											
/feldspathic											
Sapphire									0.0023		
Scheelite	CaO4W	5.8-6.2	4.5-5		10.2	5.75	3.9	E	NM		
Schist/ hornblende											
/ mica				25.85							
Senarmontite	O3Sb2	5.3	2						0.019		
Serpentine	H4Mg3O8Si2	2.5-2.6	4			11.48	2.7	P	0.14 to 0.016		
Shale		2.62		18.0/10.9							
Siderite	FeCO3	3.7-3.9	3.5-4.5			6.78	3.2	E	0.743 to 0.16	1 to 4.3 E-3	1.2 E-6
Silica/synthetic					1-2						
Silica sand		2.65		18.15							
Silicon carbide		2.73		28.8		2.5	E				
Sillimanite						NC			M		
Silver/ native		9.6-12	2.5-3			> 81					
/pure		10.5									
/ore		2.72		19.03		> 81	3.0	E			
Silver chloride	AgCl				4 Ag						
Silver iodide	AgI				5-6						
Silver sulfide	AgS				10 Ag						
Sinter		3.0		9.68							
Slag		2.93		17.38							
Slate		2.48		15.2							
Smaltite	As2Co	5.7-6.8	5.5-6			> 81	2.9	E			
Smithsonite	CO3Zn	4.3-4.5	5			5.02	5.6	N	0.029		
Sodium Silicate		2.1		14.3							
Sphalerite	ZnS	3.9-4.2	3.5-4		2-2.3	5.29	3.9	N	0.182 to -0.0004/NM	-13 E-6 to 427 E-6	-3.2 E-9 to 105 E-9
Sphene						NC			NM		
Spinel	Al2MgO4	3.5-3.7	8			6.77			0.001		
Spodumene ore	AlLiO6Si2	2.75		15.07							
Stannous oxide/synthetic					6.6						
Staurolite						NC			M		
Stibnite	Sb2S3	4.5-4.6	2			11.15	3.1	E	0.022 to 0.013		
Strengite	FeH4O6P				2.8						
Strontium H apatite					8						
Sulfides/ heavy		3.56		12.54							
Sulfur	S	2	2	12.65	2	7.03	4.9	P	NM		
Syenite/ augite		2.73		16.4							
Sylvanite	(AuAg)Te2	7.9-8.3	1.5-2								
Sylvite	KCl	2.0	1.5-2			4.7					
Taconite		3.37		17.9/16.4							
Talc	H2Mg3O16Si4	2.7 to 2.8	1		1.9	9.41	3.0	E	0.026		
Tetrahedrite	Cu8S7Sb2	4.4-5.1	3-4.5						0.08		
Thorium dioxide	O2Th				9-9.3						
Tin ore		3.94		11.88							
Titanium ore		4.23		13.2							
Topaz	AlF2H22SiO6	3.5-3.6	8			6.09	5.6	P	-0.0006		

Tourmaline	Al3B2HgH2O21Si4	3-3.2	7-7.5		4	5.17	3.2	N	0.0012/WM	
Tremolite	CaMg3O12Si4	2.9-3.4	5-6		ND					
Tungsten oxide	O3W	7.16			0.5					
Uraninite	UO2	10.3-10.6	5-6		3-4					30 E-3
/synthetic					3.5-6					
U3O8	O8U3				4					
Uranium ore		2.7		19.7						
Vanadinite	ClO12Pb5V3	6.8-7.1	3							
Vermiculite	(AlFe)2Mg3O12Si3	2.7	1.5		2.3					
Willemite	O4SiZn2	3.9-4.2	5.5			5.55			0.076/WM	2.4 E-3
Witherite	BaCO3	4.2-4.3	3-3.5			5.42			0.0064	
Wolframite	(Fe/Mn)WO4	7.1-7.5	5-5.5			12.51(C)	3.3	E	0.1	
Wool					2.2					
Wulfenite	MoO4Pb	6.8	3			6.29	5.3	E		
Xenotime						NC			M	
Yttrium oxide	O3Y2				9					
Zinc ore		3.68		13.66						
Zinc oxide, see Zincite										
Zincite	OZr	5.4-5.7	4.5-5			approx 34			0.038	
/synthetic		5.6			9-10					
Zirconium oxide	O2Zr				4					
/synthetic					10-11					
Zircon (beach sands)						6.09	5.0	P		
Zircon	O4SiZr	4.2-4.7	7.5		5.8	NC	5.3	N	0.134 to 0.0002/NM	

charge) when the surface charge is zero. (For silver iodide, the ions causing the surface charge is silver, Ag^+, and **not** H^+; thus, the zpc condition is given as pAg and not pH.) For concentrations that are more acid than these conditions, the surface charge is positive; for concentrations that are more basic, negative. Thus, for a particle with zpc of 5.26, for pH of <5.26, the surface charge is positive; >5.26, the charge is negative. So what? For froth flotation, if the mineral has a zpc of 5 and the gangue has a zpc of 10, at an intermediate pH the mineral would have a negative surface charge, the gangue would have a positive surface charge. Thus, we could have a "collector" which would preferentially adsorb to the mineral surface to make that surface hydrophobic, a "depressant" to preferentially adsorb to the surface of the gangue to make it hydrophillic and viola; we might have a flotation system that would work. This is a simplistic outline of the use of zpc to identify *potential* for flotation. Look for differences in zpc between the target species.

Another "so what?" is to guide us in coagulation and flocculation. The zpc for oil dispersed in water is about 2. Thus, if the water has a pH of 8 (as occurs in many Industrial Waste Water treatment facilities), then the oil drops will have a negative charge; they will be stable and not want to coalesce and form larger drops. Thus, we know that we will have to add chemicals that will adsorb to the oil surface and neutralize the charge. Thus, for small diameter particulates, look for the degree to which the zpc deviates from current fluid conditions. The greater the difference, the more processing effort will be needed to control and manage the surface charge. In this column 6, ND means that no charge could be detected in the whole range of pH.

Some particles acquire an electrostatic charge when placed in an applied voltage gradient. To determine if electrostatic separation of solids from solids is feasible, we could use the relative permittivity of particles (in column 7). The relative permittivity of water is 80. A useful cut-off value between conducting and non-conducting particles is 10.

The shorthand terminology used is:

NC when the permittivity is <10
C when the permittivity is >11.

If both particles conduct, they still can be separated if the "threshold voltage gradients" at which they acquire a charge are different. Threshold gradient information is given in column 8. The sign of the charge acquired compared to the sign of the "active electrode" is summarized in column 9, where:

P means the particle acquires a **positive** charge in the field of a *negative* active electrode and **no** charge in the field of a *positive* active electrode.
N means the particle acquires a **negative** charge in the field of a *positive* active electrode and **no** charge in the field of a *negative* active electrode.
E means the particle acquires a charge in the field of either a positive or a negative active electrode.

Thus, use the data in these three columns, 7, 8, and 9, to note differences in chargeability, differences in threshold voltages, and/or differences in response to the electrode.

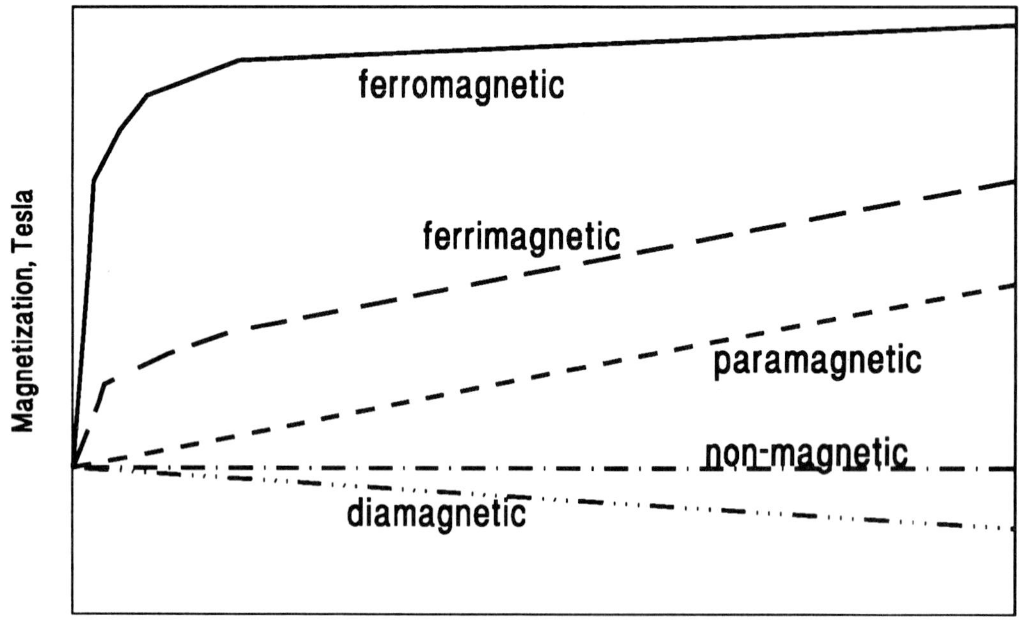

Figure C4–1 Response of Solids to an Applied Magnetic Field

Perhaps magnetism can be used to separate target species. Five possible responses to an applied magnetic field are illustrated in Figure C4–1: ferromagnetic, ferrimagnetic, paramagnetic, nonmagnetic, and diamagnetic. Since some elements have magnetic characteristics, we can consult the periodic table in Part C–5. The symbols are:

- f ferromagnetic elements (Fe, Co, Ni)
- p paramagnetic elements and compounds (Cr)
- α paramagnetic elements but form diamagnetic compounds (Ba)
- β elements may be diamagnetic but form paramagnetic compounds (Cu).

Other information is given in Table C–4. Column 10 gives magnetism *relative to iron = 100.* A negative sign indicates that the material is diamagnetic. A symbolic interpretation is

- M magnetic with relative value >10
- MM moderately magnetic with relative value between 1 and 9.99
- WM weakly magnetic with a relative value between 0.025 and 0.99
- SM slightly magnetic with a relative value between 0 to 0.024

Some separations require a value of the magnetization for the target species. This is the product of the rationalized magnetic susceptibility of the target species, κ, and the applied magnetic field **H.** Columns 11 and 12 give values of the rationalized magnetic susceptibilities, expressed in the volume and mass units respectively.

REFERENCES

DOBBY, G., and J. A. FINCH. 1977. "Capture of Mineral Particles in a High Gradient Magnetic Field." *Powder Technology* **17:**73–82.

GILLET, G. 1979. "A Cleaning and Concentration Study of Some Mineral Associates: Comparison with Other Beneficiation Processes" in "Industrial Applications of Magnetic Separation," Proceedings of an International Conference at Franklin Pierce College, Rindge, New Hampshire, July 1978, Y. A. Liu, ed. New York: Institute of Electrical and Electronic Engineers, New York, pp. 49–54.

JOHNSON, H. B. 1939. "Selective Electrostatic Separation." *Trans Am Inst Mining and Met Engineers* **134:**409–423.

KELLY, E. G., and D. J. SPOTTISWOOD. 1982. *Introduction to Mineral Processing.* New York: J. Wiley and Sons (mineral composition, hardness, density, work index)

LAWVER, J. E., and D. M. HOPSTOCK. 1974. "Wet Separation of Weakly Magnetic Minerals." *Mineral Sci Engng* **6,** 3:154–172.

OBERTEUFFER, J. A., et al. 1981. "Magnetic Separation." *Encyclopedia of Chemical Technology* **14:**709.

OGLESBY, S., JR., and G. B. NICHOLS. 1978. *Electrostatic Precipitation.* New York: Marcel Dekker, Inc.

TAGGART, A. F. 1954. *Handbook of Mineral Dressing.* New York: J. Wiley and Sons.

TU, T. Y., and C. H. YAN. 1979. "A Combined Beneficiation Method for the Treatment of Refractory Eluvial Tin Ores" in *Mineral Processing,* ed. J. Laskowski. Amsterdam: Elsevier Scientific Publishers, pp. 1146–1167.

WOODS, D. R. 1990. *Colloids, Surfaces and Unit Operations.* Hamilton, ON: McMaster University. (data on zpc)

The periodic Table of Elements is given in Table C–5.

A useful collection of data about gaseous and gas liquid systems is given in Figures C–6 and C–9. Other properties about liquids and gases are given in Part D.

PERIODIC TABLE OF THE ELEMENTS

Table of Selected Radioactive Isotopes

Nuclide	Mass	Half-life / decay	Nuclide	Mass	Half-life / decay	Nuclide	Mass	Half-life / decay	Nuclide	Mass	Half-life / decay
$_{1}$H	3	(12.3 y) β^-	$_{28}$Ni	59	(8x10^4 y) EC	$_{47}$Ag	108	(127 y) EC		205	(3x10^7 y) EC
$_{4}$Be	7	(53 d) EC		63	(92 y) β^-		110	(252 d) β^-		210	(22.3 y) β^-, α
	10	(1.6x10^6 y)β^-	$_{29}$Cu	64	(12.7 h) β^-, β^+, EC		111	(7.5 d) β^-	$_{86}$Rn	222	(3.83 d) α
$_{6}$C	11	(20.4 min) β^+	$_{30}$Zn	65	(244 d) β^+, EC	$_{49}$In	114	(49.5 d) IT	$_{88}$Ra	226	(1.6x10^3 y) α
	14	(3070 y) β^-	$_{31}$Ga	67	(78.2 h) EC	$_{50}$Sn	121	(76 y) β^-	$_{89}$Ac	227	(21.77 y) β^-
$_{11}$Na	22	(2.602 y) β^+, EC		72	(14.1 h) β^-	$_{51}$Sb	124	(60 d) β^-	$_{90}$Th	228	(1.913 y) α
	24	(15.02 h) β^-	$_{32}$Ge	68	(275 d) EC		125	(2.7 y) β^-		230	(7.7x10^4 y) α
$_{15}$P	32	(14.3 d) β^-	$_{33}$As	73	(80.3 d) EC	$_{52}$Te	127	(109 d) IT		232	(1.4x10^{10} y) α
$_{16}$S	35	(87.2 d) β^-		74	(17.9 d) β^-, β^+, EC	$_{53}$I	129	(1.6x10^7 y) β^-	$_{91}$Pa	231	(3.3x10^4 y) α
$_{17}$Cl	36	(3x10^5 y) β^-	$_{34}$Se	75	(118 d) β^-		131	(8.04 d) β^-	$_{92}$U	233	(1.59x10^5 y) α
	38	(37.2 min) β^-		79	(7x10^4 y) β^-	$_{55}$Cs	134	(2 y) β^-		234	(2.44x10^5 y) α
$_{19}$K	40	(1.3x10^9 y) EC	$_{35}$Br	82	(35 h) β^-		135	(2.9x10^6 y) β^-		235	(7.04x10^8 y) α
	42	(12.4 h) β^-	$_{37}$Rb	86	(18.7 d) β^-		137	(30.2 y) β^-		236	(2.34x10^7 y) α
$_{20}$Ca	45	(165 d) β^-		87	(5x10^{11} y) β^-	$_{56}$Ba	140	(12.8 d) β^-		238	(4.47x10^9 y) α
$_{21}$Sc	46	(83.8 d) β^-	$_{38}$Sr	90	(28.8 y) β^-	$_{57}$La	137	(6x10^4 y) EC	$_{94}$Pu	238	(87.75 y) α
$_{24}$Cr	51	(27.7 d) EC	$_{39}$Y	88	(107 d) β^+, EC		140	(40.3 h) β^-		239	(2.4x10^4 y) α
$_{26}$Fe	59	(44.6 d) β^-	$_{40}$Zr	93	(1.5x10^6 y) β^-	$_{58}$Ce	144	(284 d) β^-		240	(6.54x10^3 y) α
$_{27}$Co	56	(78.8 d) β^+, EC		95	(64 d) β^-	$_{60}$Nd	147	(11 d) β^-		242	(3.8x10^5 y) α
	57	(270 d) EC	$_{42}$Mo	99	(66 h) β^-	$_{61}$Pm	145	(18 y) EC		244	(8.3x10^7 y) α
	58	(71.3 d) β^+, EC	$_{46}$Pd	103	(17 d) EC		147	(2.6 y) β^-	$_{95}$Am	241	(432 y) α
	60	(5.27 y) β^-		107	(7x10^6 y) β^-	$_{82}$Pb	202	(3x10^5 y) EC		243	(7.4x10^3 y) α

Selected Radioactive Isotopes

Half lives are in parentheses where s, min, d and y stands for seconds, minutes, days and years respectively. The symbols for the mode of decay and resulting radiation are defined as follows:

- α alpha particle emission
- β^- beta particle (electron) emission
- β^+ positron emission
- EC orbital electron capture
- IT isomeric transition from upper to lower isomeric state
- SF spontaneous fission

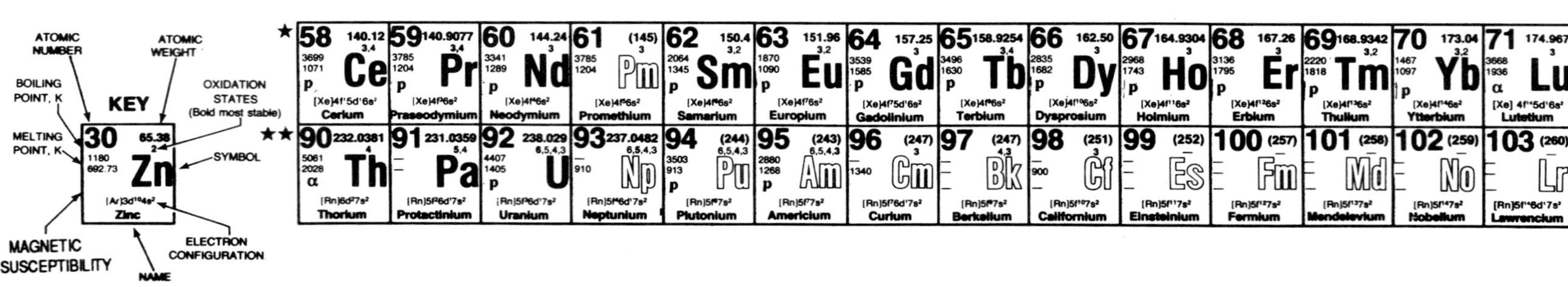

KEY: ATOMIC NUMBER; ATOMIC WEIGHT; BOILING POINT, K; MELTING POINT, K; OXIDATION STATES (Bold most stable); SYMBOL; MAGNETIC SUSCEPTIBILITY; ELECTRON CONFIGURATION; NAME

Code for magnetic susceptibility:

f ferromagnetic elements that form ferro or paramagnetic compounds.
p paramagnetic elements that form paramagnetic compounds.
α paramagnetic elements only; compounds containing these elements are diamagnetic.
β paramagnetic compounds formed; elements often are diamagnetic.

SARGENT-WELCH SCIENTIFIC COMPANY
P.O. Box 1026, SKOKIE, ILLINOIS 60076-8026

Table C–5 Periodic Table of the Elements (reprinted courtesy of Sargent-Welch Scientific Company)

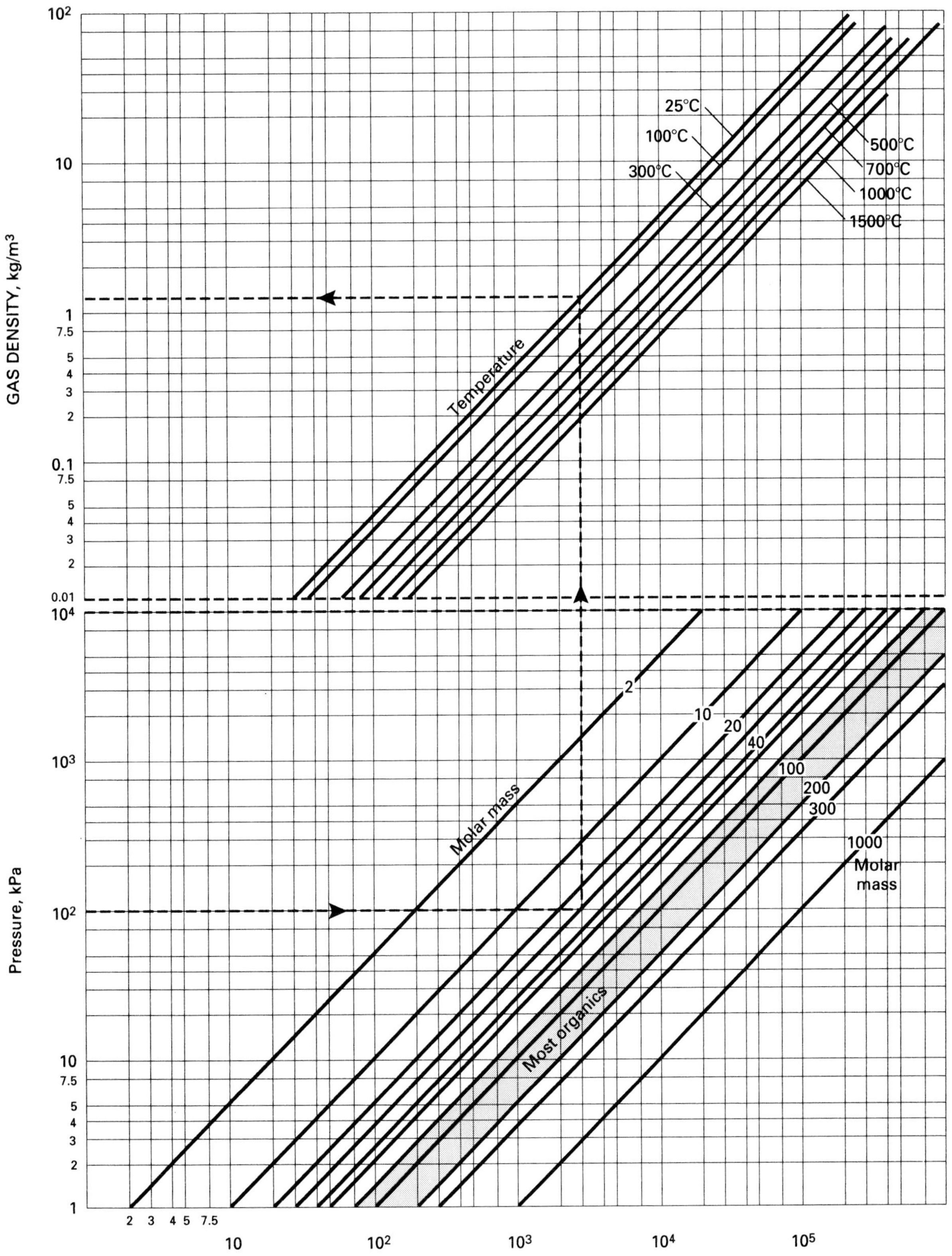

Figure C–6 Estimating Gas Density (Based on the Ideal Gas Law)

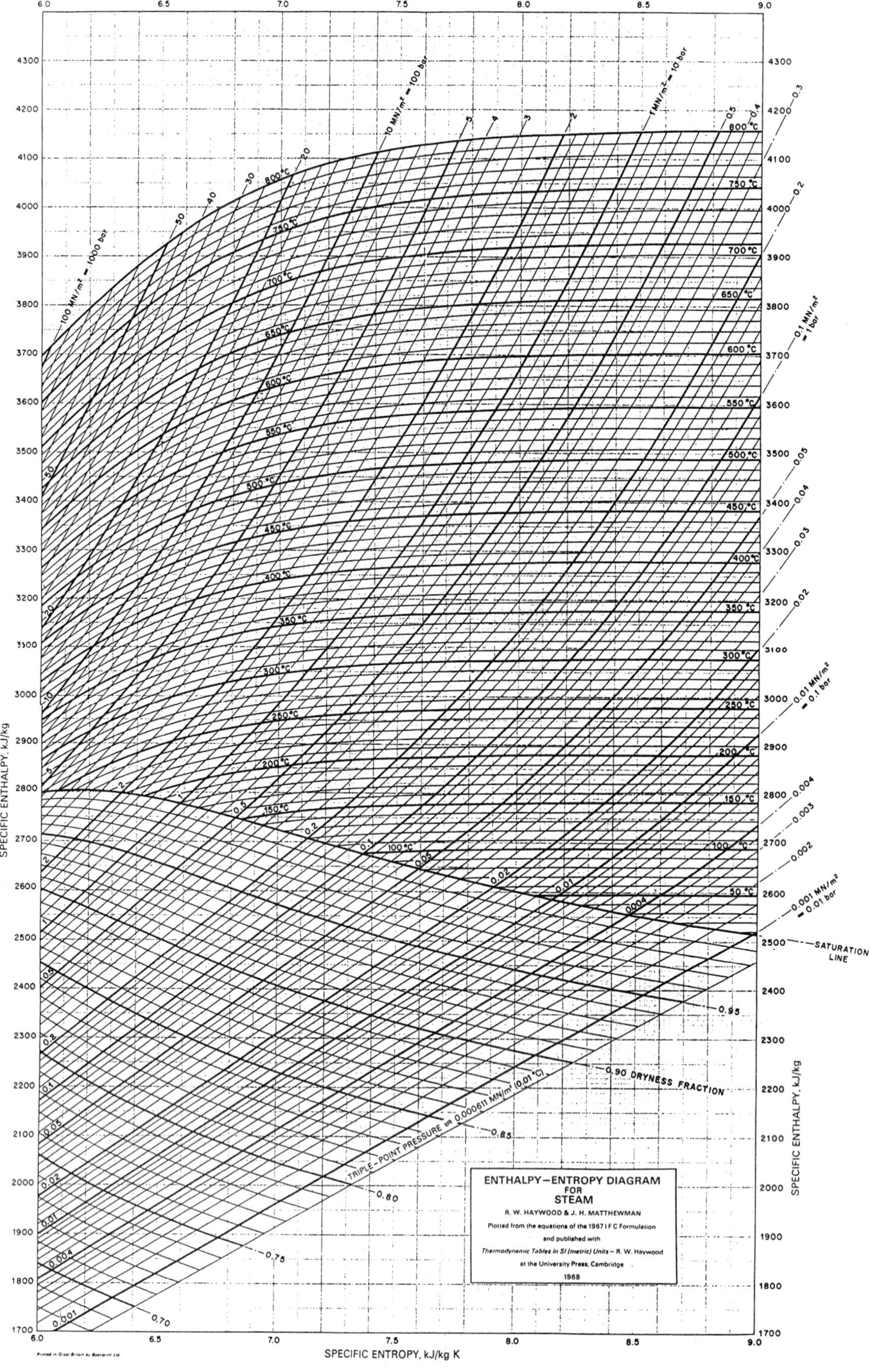

Figure C–7 Entropy-Enthalpy Diagram for Steam (reprinted courtesy of Cambridge University Press)

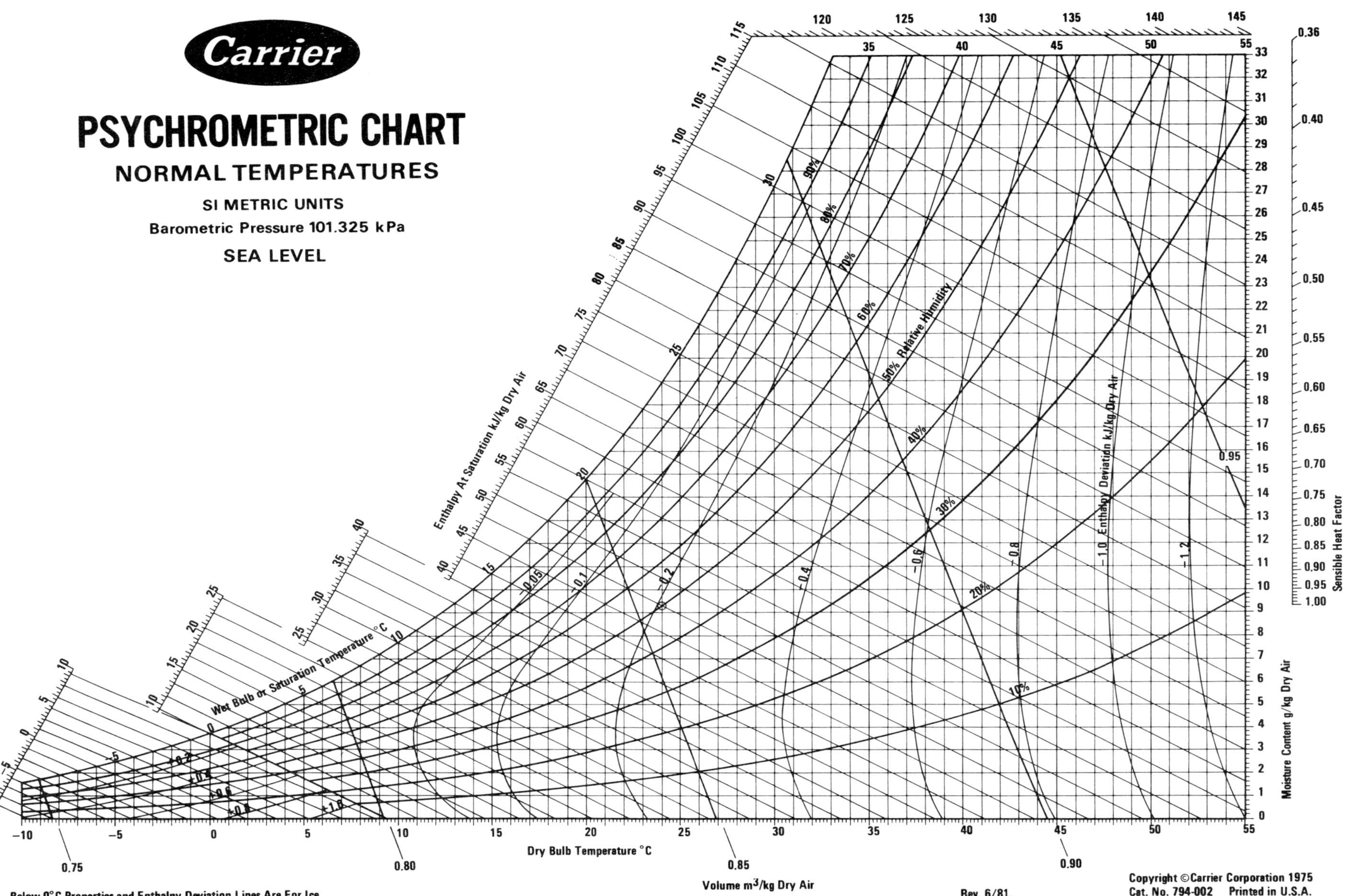

Figure C–8 Psychometric Charts a) –10 to 55°C (reprinted courtesy of Carrier Corporation, 1975)

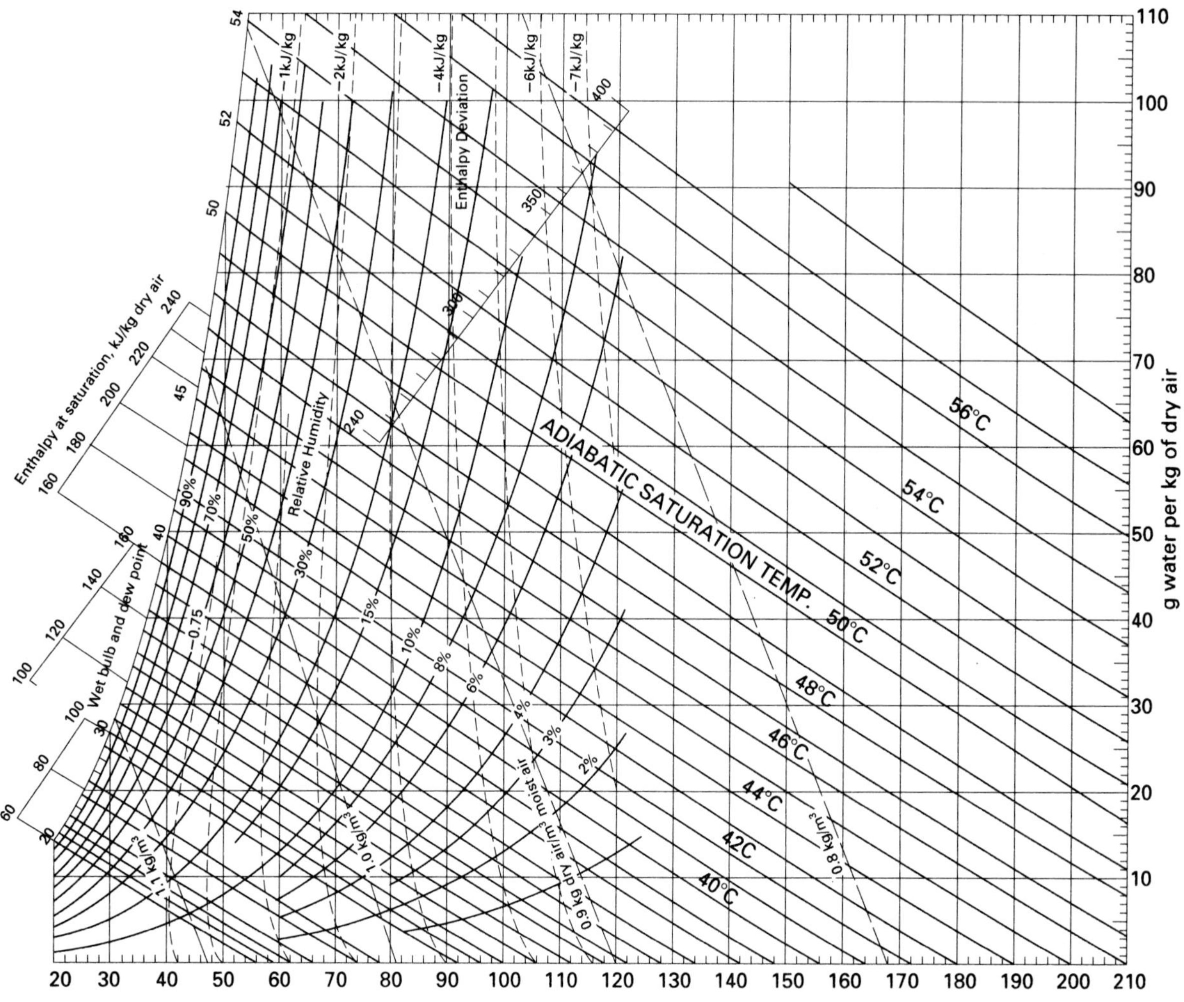

DRY BULB TEMPERATURE °C

Figure C–8 b) 20 to 210°C

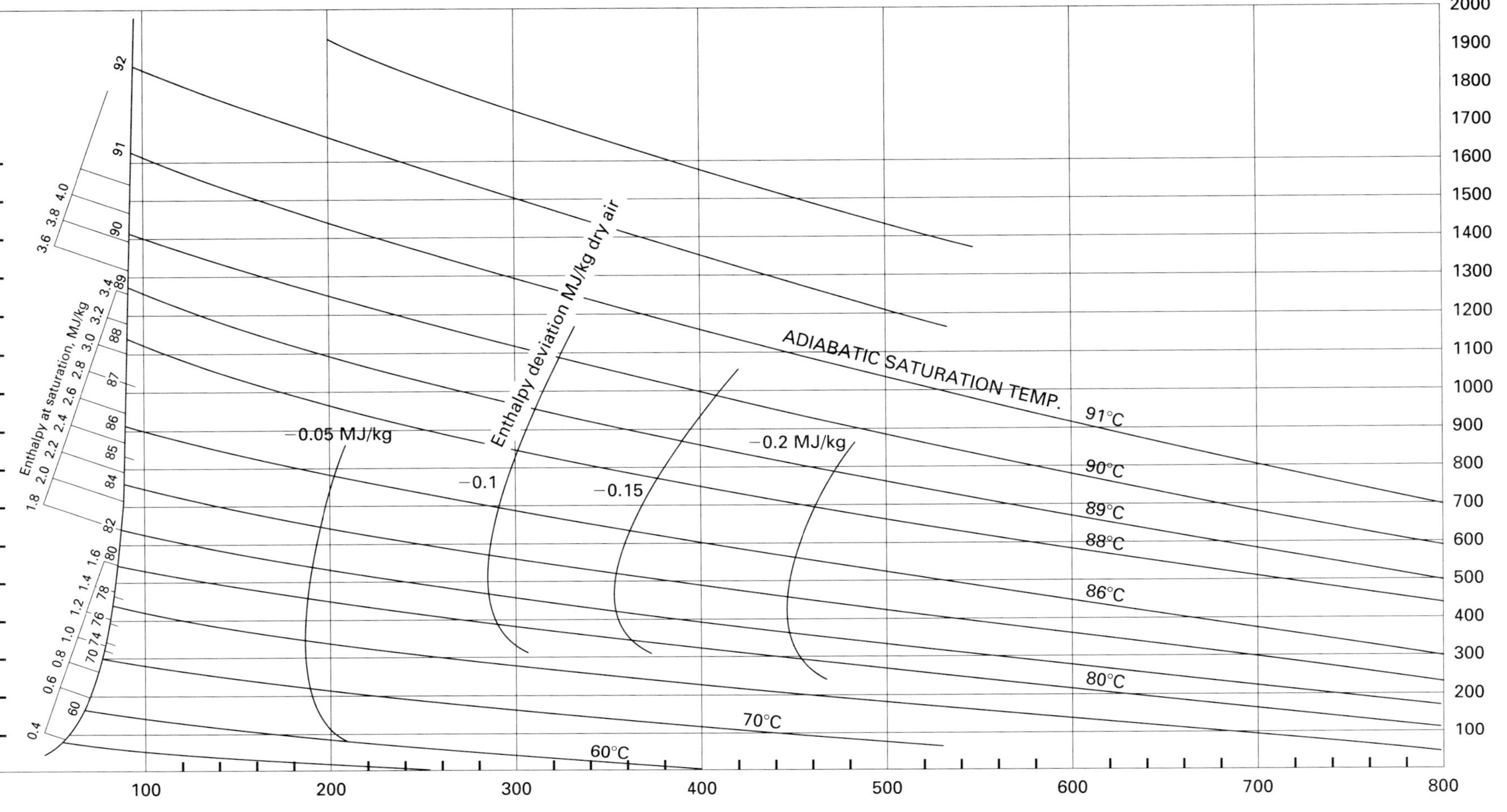

Figure C–8 c) 0 to 800°C

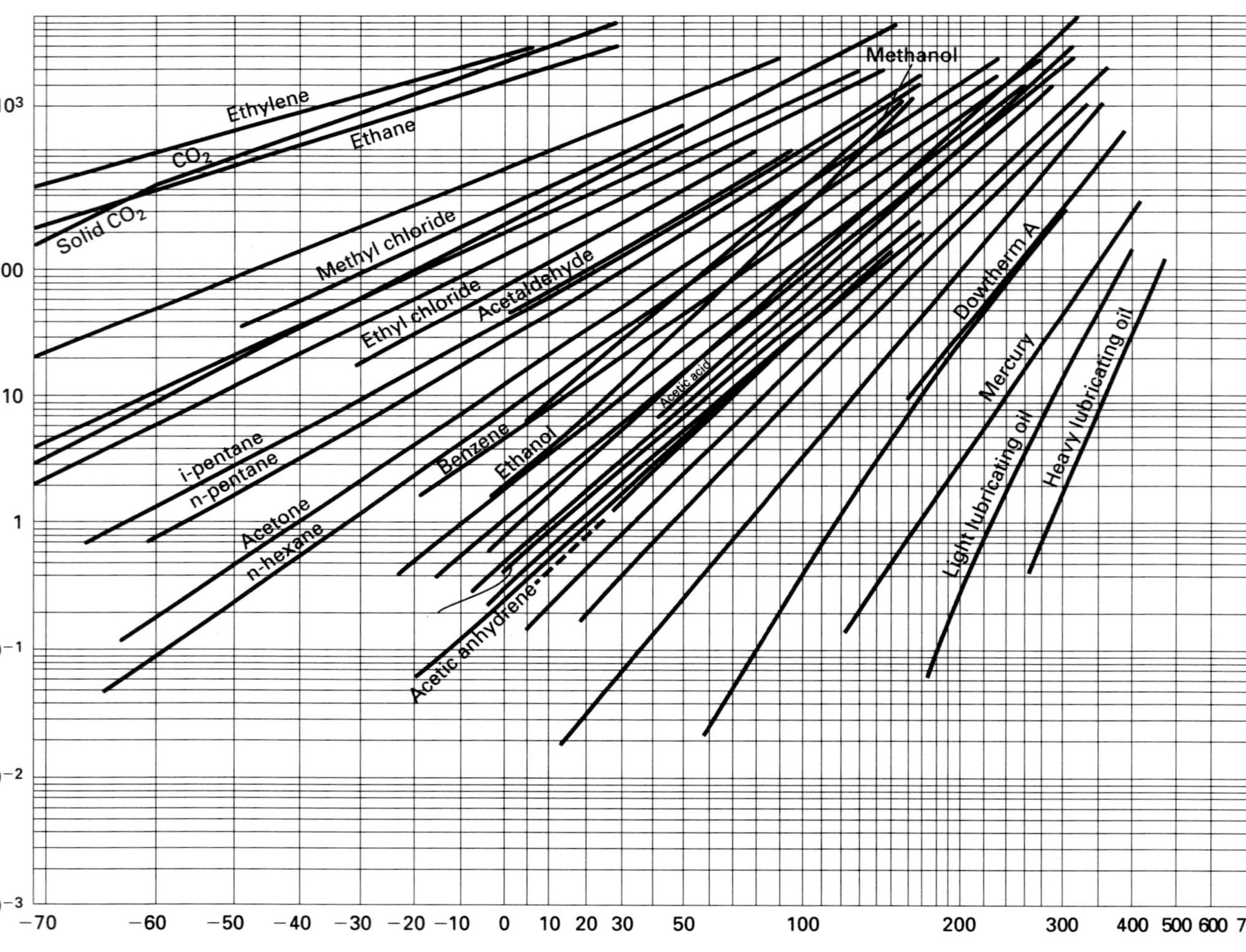

Figure C–9 Vapor Pressures of Some Compounds a) –70 to 700°C

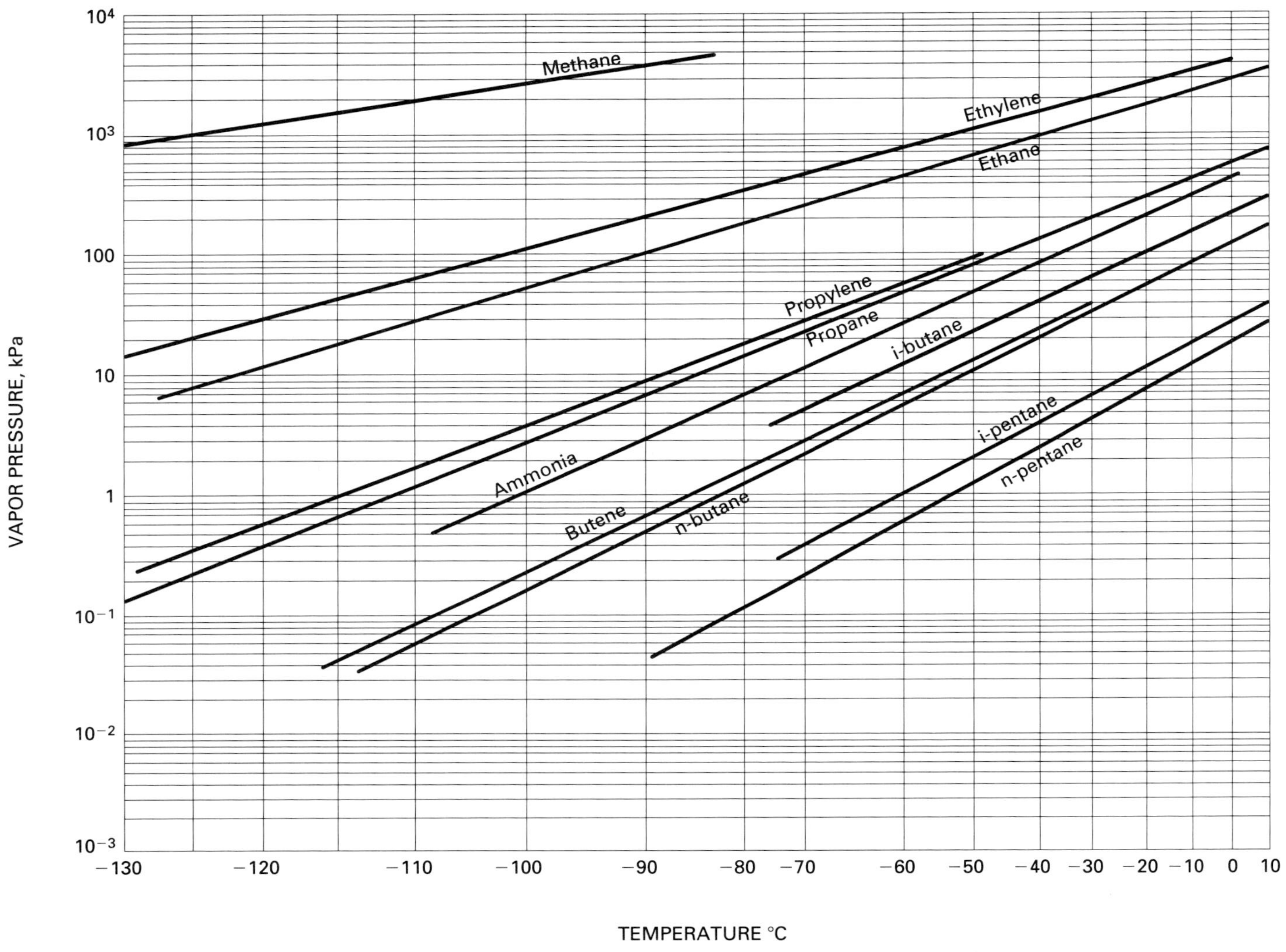

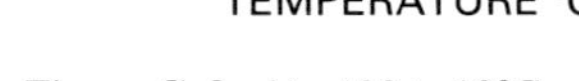

Figure C–9 b) −130 to 10°C

PART D: SELECTED PROPERTIES FOR LIQUIDS AND GASES FOR

A. Selecting Separation Processes
B. Sizing Equipment

A. SELECTED PROPERTIES FOR SELECTING SEPARATION PROCESSES

For separating species starting from a homogeneous phase, what are the key properties for us to consider?

- Molar mass: many different options depend on a difference in this.
- Something to tell us about the stability and corrosiveness of the species. This is needed when we have a multicomponent mixture to separate. One heuristic is to separate the hazardous and corrosive material first. Thus, we need some indication of this. We use the National Fire Protection Association (or NFPA) ratings. These go from 0 to 4, with 0 meaning harmless and 4 meaning extremely hazardous. Three different components are reported: Hazard to Health (the first number in the table), Flammability Hazard, and Reactivity or Stability hazard. It is the third number that is most useful at this time. The other two, namely Health and Flammability, help us to chose from competing agents (say, solvents for solvent extraction—we would chose one with small values of the H and F rating). These also are important when we look at the system, integration of parts of the system, and control.
- the Dow Material Factor is reported next. This is more for completeness than for pertinence *at this time.* This value is a weighting factor used by Dow in a comprehensive safety analysis approach that they have published. The focus is mainly on flammability; the higher the number the more hazardous the location of the plant is. Hence, this value would be used (together with the NFPA values for Health and Flammability) when we are selecting among difference "agents" and for designing the system—as opposed to when we are posing the separation options. Some of the implications of the "total Dow factor" for a section of the plant are suggested in Figure 4–7d of **PDEP**.
- the ability for the basic three materials of construction to withstand the chemical is given next. The coding is patterned after the NFPA ratings and is a scale from 0 to 4. Zero means <0.05 mm/a of corrosion (<2 mils per year) whereas 4 means >3 mm/a (>120 mpy). Great care was taken in developing these ratings; great care is needed in interpreting them as well. The ratings refer more to *potential for corrosion under probable separation conditions.* What does this mean? Usually the mixture being processed may contain trace amounts of water and will probably be at 100 to 200°C operating conditions or at the boiling point. These are not the ratings of "pure" species at room temperature. Thus, carbon steel could be quite appropriate for a storage tank for the final product whereas it would be inappropriate for a distillation column. In general, I have tried to rate these more on a "worst case" scenario without being too extreme. The values are meant as a guide for selecting separation sequencing only. The three materials chosen were: copper, carbon steel (mild steel), and 316 stainless steel (18-8-2). These were chosen because they are economical first choices and because they modestly cover oxidizing and reducing conditions. More details about Materials of Construction are given in Appendix I of **PDEP**.
- the melting or fusion temperature is given next, since this helps us see if melt crystallization is possible.
- the normal boiling temperature, at atmospheric pressure, helps us see the appropriateness of distillation and condensation as possible options.

Many of the options require a matching of or a difference in the solubilities of the species. This is "measured" by a combination of:

- the Molar volume;
- the Hildebrand Solubility parameter: polar contribution; and
- the Hildebrand Solubility parameter: hydrogen bond contribution.

These three parameters are given in the next three columns. These are important for the selection of absorption, adsorption, solvent extraction, and most of the membrane processes. The molar volume is the molar mass divided by the liquid density. The total Hildebrand solubility parameter is:

$$\delta^2 = \delta_d^2 + \delta_p^2 + \delta_h^2 \qquad \text{(D-1)}$$

where: δ = Total Hildebrand solubility parameter
δ_d = dispersion contribution (which is about constant)
δ_p = polar contribution (which makes a difference)
δ_h = hydrogen bond contribution (which makes a difference).

The dissociation constant pK_a helps us to see if the species dissociates in water. Many of the separation options depend on the target species being ions (ion exchange, reverse osmosis, electrodialysis). Furthermore, many of the materials of construction corrode in the presence of weak or strong acids or bases. The dissociation constant gives us an idea about the potential for corrosion. If the pK_a is large, then few ions are present and those options that depend on ions being present are not viable. On the other hand, the materials of construction may not corrode.

The solubility product, pK_{sp}, indicates whether solution crystallization or precipitation have potential.

For biological degradation of species, the important measures are the rate of COD removed and the BOD/COD ratio. If the former is not large and the latter is not greater than 0.5, then micro-organisms cannot biodegrade the chemical species. Thus, biological reaction is not an option. Data are scarce and dependent on the biological conditions used. The data cited here come from two prime sources: activated sludge conditions (from Pitter (1976) and Despand (1987) and static, culture flask studies (from Tabak et al., 1981). The former are quantitative results; the latter use the following code describing the species performance relative to that of *phenol:*

D: significant degradation with rapid adaptation of the microorganisms;
A: significant degradation with gradual adaptation;
B: slow to moderate biodegradation with significant volatization;
C: very slow biodegradation requiring lengthy adaptation;
N: negligible biodegradation;
T: significant initial degradation followed by toxicity.

Thus, species with N,T,C, and perhaps B could not be separated by biological reaction.

Since energy is often an important consideration for options, the heat of fusion and of vaporization are given.

For oils, some of the key properties are:

IV: the iodine value. This is a measure of the double bonds or an approximate measure of the degree of unsaturation of the oil (fat or fatty acid). [Defined as the number of g of iodine absorbed by 100 g of oil, fat, or fatty acid under standard conditions.]

SV: the saponification value. This is a measure of the amount of base required to make soap. [Defined as the mg of KOH required to saponify 1 g of oil, fat, or fatty acid. The reaction of the fat with alkali produces glycerol and the salt or soap of the alkali metal.]

AV: the acid value. This is a measure of the degree of hydrolysis that has occurred in the oil or fat. [Defined as the mg of KOH required to neutralize the free fatty acids in 1 mg of oil or fat. The triglycerides of oils and fats hydrolyze to give glycerol and free fatty acids. This is a measure of the fatty acids present.]

For oils, values for these three measures are reported in Table Da.

All the properties are given in SI units at 25°C even if the material does not exist at that condition in the form suggested.

The greatest challenge in compiling these data was terminology. In general, I tried to use the terminology used by Danbert and Danner (1985) and updates since this is the follow-up resource we would probably use for more accurate data. To help locate the appropriate entry, consult the Chemical Index first. This gives synonyms, cross references, and the term used in this book. For example, 1,2-Dichloroethane is listed under the E's as Ethylene dichloride, whereas 1,1-Dichloroethane is listed under the D's as that name.

The same terminology is used throughout the book; Table **Db** gives all of the same entries that are given in Table D**a**.

B. SIZING EQUIPMENT

Order-of-magnitude sizing can lose much of its accuracy if *reasonable* values are not used for the thermal and physical properties of the species. However, this assumes rapid access to such data. The purpose of Table Db is to provide such data and, where available, to provide sources for more accurate values.

The tables list the species in the same order as listed in Table D**a**. To facilitate use, only the value of the pure component property at 25°C, 101 kPa is given, together with some indication of the variation in temperature. By limiting the amount of information given, all of the properties are given in the table so that one can easily scan the columns and see trends. Thus, scanning the column for the thermal conductivity of liquids, we see that the value is usually about 100 to 200 mW/m.K. We also note that hydrogen and water are exceptions. Hence, even though the table entries are blank for some species, we can estimate approximate values by locating similar compounds. Furthermore, the entries are for pure species. We usually deal with mixture. Hence, we need the properties of mixtures. Again, a quick, scan of the table will allow us to approximate the mixture values.

Such tabulation of data is dangerous and misleading. Most of the data for both the liquid and gas states are given at 25°C, one set of data has to be extrapolated. Extrapolation is dangerous. Whenever the extrapolation exceeded 100°C the table entry is bracketed. For all of the properties except *liquid viscosity,* a linear variation with temperature is used. This is reasonable for most properties except for density near the boiling point. The "general" variation in temperature for the different properties is illustrated in Figure Db–1. For sizing condensers and reboilers, the properties at the boiling temperature are needed. For selected species, these are reported as*. The variation with "boiling temperature" is reported for some of these; this really represents a variation in pressure for these properties.

Table Da-1: Selected Properties for Selecting Separation Processes (at 25°C unless noted)

	mol mass	NFPA H F S	Do	Cuc s s s	freezing temp. °C	boiling temp C	Molar volume	δ_p $(J/cm^3)^{0.5}$	δ_h	pKa	pKsp	Biodegr mg COD h.g	BOD/ COD	ΔH_m kJ/kg	ΔH_v kJ/kg
Abietic acid:	302.46				172										
Acenaphthene:	154.21				93-5	279									
Acetal:	118.17	2 3 0			102.2	143.2									277
Acetaldehyde:	44.05	2 4 2	24	0 0 0	-125.0	20.8	55.9	10.6	12.7					55	570
Acetamide:	59.07			0 3 0	82.3	221.2	51			0.63				266	951
Acetanilide:	135.17	3 1 -		1 1 1	114	304	110			0.4 (+1)		14.7			
Acetic acid:	60.06	3 2 1	10	4 3 2	16.6	117.9	57.2	12.2	18.9	4.76				195	405
Acetic anhydride:	102.09	2 2 1		1 0 0	-73	139.6	94.4	11.3	10.2					103	385
Acetone:	58.08	1 3 0	16	? 0 0	-94.35	56.2	73.5	9.8	11				0.39	98	517
Acetone cyanohydrin:	85.11			1 1 1	-19		91.3								
Acetonitrile:	41.05	2 3 1	24	1 0 0	-48	81.6	52.5	11.1	19.6					199	727
Acetophenone:	120.15	1 2 0		0 0 0	20.5	202.6	116.9	11.9	8.2						322
1-Acetoxy-1,3,-butadiene:	112.13						118.7	9.9	8.7						
Acetylacetone:	100.12	2 2 0			-23	140.4	102.7	11.8	10.7						
Acetyl chloride:	78.50	3 3 2	16	0 4 2	-112	52	71.1	10.6	3.9						330
Acetylene: (Ethyne)	26.04	1 4 3	40	4 0 0	-80.8	-84	28.7								96
Acetyl peroxide:	118	1 2 4	40		30	63									
Acetylsalicylic acid:	180.16	1 1 0		? 1 1	138					3.48					
Acrolein: (2-propenal).	56.06	3 3 2	24	1 1 1	-86.9	53 66.7	11.1	12.2				D			506
Acrylamide:	71.08				84.5	....	63								
Acrylic acid:	72.06	3 2 2		? 0 0	13	141.6	68.6	12.8	17.9	4.25				155	517
Acrylonitrile:	53.06	4 3 2		0 0 0	-83	77	65.8	12.5	14			D	0	125	614
Adipic acid:	146.14	- 1 -		? 0 1	153	265	107			4.43				114	552
Adiponitrile:	108.14	4 2 -			1	295	113.7								
Air:	29	...													
Allyl acetate:	100.12				6	103	107.9	9.5	8.3						
Allyl acetoacetate:	142.16				75	194	137.1	11.3	10.5						
Allyl alcohol:	58.08	3 3 1		0 ? 0	-129	97.1	68	11.8	18.7	15.5					689
Allyl amine:	57.1				-88	53	75			9.49					
Allyl bromide:	120.98	3 3 1			-119	70	86.5								
Allyl chloride:	76.53	3 3 1		1 3 1	-134.5	45	81.6	8.9	7.3						380
Allyl chloroformate:	102.54					27									
Allyl cyanide:	67.09				-87	116-21	80.4	11.8	13.2						
Allyl iodide:	167.98				18	102	91.4								
Aluminum hydroxide:	78.00			1 1 1	300						15.				
Aluminum sulfate:	342.15			3 3 3	770(dec)										
2-Aminobiphenyl:	169.23	2 1 0			50-3	299				3.8(+1)					
2-Amino-1-butanol:	89	2 2 0			-2	174									
Aminodiphenylamine:	184				77										
12-Aminododecanoic acid:	215				185										
2-(2-Aminoethoxy)ethanol:	105.14					218									
N-(2-Aminoethyl)ethanolamine:	104	2 1 0				242	101	12.8	20.8						
4-(2_Aminoethyl)morpholine:	130	2 2 0				201									
N-(3-Aminoethyl)piperazine:	129	2 2 0				218	132	11.7	9						
Aminoheptane	115.22				-	142									
6-Amino-1-hexanol	117.19				57										
2-Aminonaphthalene	143.19				112	306									
3-Amino-1-propanol	75.1				10	187									
4-(3-Aminopropyl)morpholine:	144.2				-15	224	146	10.2	7.3						
Ammonia:	17.03	3 1 0		0 0 0	-77	-33.4	20.8	(13)	(11)	4.75					1366
Ammonium acetate:	77.08			3 0 0	114		65.9								
Ammonium bicarbonate:	79.06			? 0 1	107.5	50.0									
chlorate:	101.49				102 exp										
chloride:	53.5				340 s	520									

hydroxide:	35.05			3 1 1	-77										
nitrate:	80.	2 1 3	29	3 0 0	169.6	210d	48								
oxalate:	142.12			1 0 1	...	...	94.7					9.3			
phosphate:	115.04				190										
sulfate:	132.14				234 d										
p-Amyl acetate:	130.2	0 1 0		0 1 0	...	...	149	7.3	5.7						316
act-Amyl alcohol:	88.2				<-70	128	108.2	9.1	13.9						503
p-n-Amyl alcohol:	88.2	1 3 0		3 ? 0	-78.5	137.9	108	9.1	13					112	
sec-Amyl alcohol:	88.2	1 3 0			...	119.5	109	8.5	13.7						
tert-Amyl alcohol:	88.2	1 3 0			-11.9	102	109	10.0	12.4						443
n-Amylamine:	87	3 3 0			-55	103	113.6								413
Amyl bromide:	151	1 3 0			-95	129.7	124								202
Amyl butyrate:	158.23				-73.2	186.4	181								
Amyl chloride:	106.6	1 3 0		1 3 0	-99	108.4	120								309
Amyl ether:	158				-69.4	186									
Amyl formate:	116	1 3 0			-73	132	128								
4 tert-Amyl phenol	164.25				88	255									
Anethole:	148				22.5	235	149							108	299
Aniline: (benzeneamine)	93.1	3 2 0	14	3 1 0	-6.2	184.4	91.5	5.1	10.2	4.63		19	0.61	113	434
Anisidine:	123				5.2	225	112			4.52					
Anisole:	108	1 2 0		1 1 1	-37.5	155	108	4.1	6.8						341
Anthracene:	178	0 1 -		1 1 1	217	340						A		162	317
Anthraquinone:	208				284	380									
Arginine:	174.2				226 d										
Argon:	39.94				-189.2	-185.7									
Azelaic acid:	188.22				109-11										
Azobenzene:	183.23				68	293									
B*															
Barium carbonate:	197.37			1 1 1		d					8.29				
chloride:	208.24				962 tr	1560									
chromate:	253.32										9.93				
fluoride:	175.33				1355	2137					5.77				
iodate:	487.14				d						8.8				
iodate- hydrate: (H2O)	505				200						9.19				
manganate:	256.27				200 d						9.6				
oxalate:	225				400 d						7.64				
oxalate: (2H2O)	261										6.92				
sulfate:	233.39			1 1 1	1580						9.89				
Benzaldehyde:	106	2 2 0	24	3 0 0	-26	179	101.5	7.4	5.3			119	0.82		403
Benzamide:	121			? 2 1	128	288									
Benzene:	78.1	2 3 0	16	3 1 1	5.5	80.1	89.1	8.6	4.1					126	393
1-3-Benzenediol:(resorcinol)	110.11	- 1 0			111	178	86.6	8.4	21.1			57.5 D		193	
Benzenethiol:	110	2 2 -			-14.9	169				6.5				104	336
1,2,3-Benzenetriol:	126.11				133	309									
Benzidine dihydrochloride:	257														
Benzoic acid:	122.12	2 1 0	4	1 3 1	122.13	249	96.5	7.0	9.8	4.2		88.5	0.75	148	415
3-4-Benzofluoranthene:	252.32				168	---									
Benzonitrile:	103.12	1 1 1			-13	190.7	102.1	9.0	3.3					106	446
Benzoperylene:	276.34				277-9	>500									
Benzophenone:	182.22				49	305									
1,2-Benzopyrene:	252.31				175	495									
Benzothiazole:	135.18				2	231	108.5								
Benzyl acetate:	150.18				-51	206									329
Benzyl alcohol:	108.14	2 1 0		4 1 3	-15.3	205.3	103.8	12.2	15.6					83	470
Benzylamine:	107.16				10	184									
Benzylaniline:	183.25				27	306									
Benzylbenzoate:	212.25				18	323									367

continued

Table Da-1: Selected Properties for Selecting Separation Processes (at 25°C unless noted)

	mol mass	NFPA H F S	Do	Cuc s s s	freezing temp. °C	boiling temp C	Molar volume	δ_p $(J/cm^3)^{0.5}$	δ_h	pKa	pKsp	Biodegr mg COD h.g	BOD/ COD	ΔH_m kJ/kg	ΔH_v kJ/kg
Benzyl n-butyl phthalate:	312	1 1 0				370	306	11.3	3.1			D			
Benzyl chloride:	126.59	2 2 1	14	0 0 0	-39	179.3	115.1	7.2	2.7						
Benzyl cyanide:	117				-23	233.5									452
Biphenyl: (Diphenyl)	154.21	2 1 0	14	0 0 0	71	255.9	178.1	1.0	2.1					121	296
Bis(2-chloroethyl) ether:	143.02	2 2 0			-43	116-7	117.6	9.0	5.7					61	316
Bisphenol A:	228.29				158-9										
Borneol:	154.24	2 2 0		1 3 1	210	sublim	152					8.9			
Bromine:	159.8	4 0 1		3 3 3	-7.2	58.8	50.7							66	188
o-Bromoaniline:(2-)					31	229									
m-Bromoaniline:(3-)					17	251									
p-Bromoaniline:(4-)_					66										
Bromobenzene:	157.01	2 2 0	14	1 1 0	-30.8	156	105	8.2	0					68	241
1-Bromobutane:	137.02				-112.4	101								49	238
2-Bromobutane:	137.02	2 3 0			-117.4	21	109.3	4.4	3.0						
Bromochloromethane:	129.39			1 1 1	-86.5	68.1	66.9	11.1	6.5			D			
Bromodichloromethane:	163.83				-57.1	90	82.7	4.1	6.1						
Bromoform:	252.73			1 1 1	8.3	149.5	87.5					A			
Bromonaphthalene:	207.07				6.2	281	139.7	3.1	4.1						473
4-Bromophenyl ether:	249.11				18.71	310.1	175.3								
1-Bromopropane:	123	2 3 0			-110	71									
2-Bromopropane:	123				-89	60									
o-Bromostyrene:	183.05				-44	86-7	130.5	9.6	0						
o-Bromotoluene:(2-)	171.04	2 2 0		0 0 0	-27.8	181.7	120.2	8.8	0						
m-Bromotoluene:(3-)	171.04				-40	184									
p-Bromotoluene:(4-)	171.04	2 2 0	14	0 0 0	28.5	184.3	122.2	8.4	0					87	
Bromotrichloromethane:	198.27				-5.6	104.7	98.5	13.8	3.1						
Bromotrifluoromethane:	148.91				---	-59		2.5	0						
Bromouracil:	190.99				>300										
1,2-Butadiene:	54.09				18-9									129	451
1,3-Butadiene:	54.09	2 4 2	29	0 0 0	-108.9	-4.4	87.1	6.3	4.5					148	419
Butadiene dioxide:	86.09				4	144	77.3	13.1	11.6						
Butadiene sulfone:					64	151								88	436
Butane: (C4)	58.12	1 4 0	21	0 0 0	-138.4	-0.5	100.4	0	0					80	386
Butanediol: (Butylene glycol)	90.12	0 2 0		1 3 1	<-50	190.5	89.9	12.2	22.2			40			
1,2-Butanediol:	90.12	0 2 0			-50	195									
1,3-Butanediol:	90.12	1 1 0		1 3 1	<-50	207.5	89.4	10	21.5						649
1,4-Butanediol:	90.12	1 1 0		1 3 1	120	235	88.6	13.6	27.0						
2,3-Butanediol:mixture															
1-Butanol:	74.12	1 3 0	16	0 2 0	-89.5	117.2	91.5	10	15.4			84		125	591
2-Butanol:	74.12				99.5	91.7	9.1	14.8				55			562
tert-Butanol:	74.12	1 3 0			25	82.9	95							92	719
1-Butene: (a-Butylene)	56.11	1 4 0	21		-185.3	-6.3	94.3	4.2	2.9					69	391
cis-2-Butene:	56.11	1 4 0	21		-138.9	3.7	90.3	3.8	5.8					130	415
trans-2-Butene: (b-Butylene)	56.11	1 4 0	21		-105.5	0.9	92.9	3.6	4.2					174	416
n-Butyl acetate:	116.26	1 3 0	16	0 3 0	-78	124-6	131.8	7.8	6.8						309
2-Butyl acetate:	116.26				-99	111-2	133.2	7.7	5.0						
Butyl acetoacetate:	158.20	1 2 0				214	163.9	9.5	10.3						
Butyl acrylate:	128.17	2 2 2	14			145	143.4	8.3	6.8						
n-Butylamine: (1-Butanamine)	73.14	2 3 0	16	0 0 0	-49.1	77.8	98.7	8.1	8.0	3.4(+1)					439
2-Butylaniline:	149.24	3 1 0			---	122-5	156.6	9.1	6.6						
n-Butyl benzene:	134.22	2 2 0			-88	183	156.1	6.5	0.0					84	293
Butyl benzoate:	178.23	1 1 0		1 1 1	-22	249	176.5	9.4	5.9						
n-Butyl cyclohexane:	140.27				-74.7	181	175.5	2.4	0.0					101	
n-Butylcyclohexylamine:	155.28					209	185.1	6.3	3						

n-Butylcyclopentane:	126.24				-108	156.7	160.9	2.9	0.0						
1-2-Butylene oxide:	72.11	3 3 2	24		-150	63.3	86.2	8.0	7.0					90	418
n-Butylethanolamine:	117.2	1 2 0				192	132	9.1	15						
Butyl ether:	130.22	2 3 0	16		-98	142-3	170.4	4.3	4.5						283
Butyl formate:	102.13	2 3 0				106	115								363
tert-Butyl hydroperoxide:					4										
n-Butyl lactate:	146.19	1 2 0		0 0 0	-43	185-7	148.6	6.5	10.2						
Butyl mercaptan:	90.188	2 3 0			-115	98									
2-Butyloctanol:	186.3	1 1 0				252	224.2	6.3	10.0					116	257
Butyl phenol:	150.22				-20	235	154.1								
Butyl propionate:	130	2 3 0			-89.5	145									300
Butyl salicylate:	194.2				-5.9	270	181.7	11.2	5.4						
Butyl stearate:	340.6				26.3	350	382	3.7	3.5					167	294
o-n-Butyltoluene:	148.2						171.3	6.1	0.0						
m-n-Butyltoluene:	148.2					192	173.7	6.0	0.0						
p-o-Butyltoluene:	148.2						174.1	5.9	0.0						
Butyl vinyl ether:	100.2	2 3 2	24			94	129.5	6.1	5.5						322
1-Butyne: (Ethylacetylene)	54				-125.7	8.1								112	
2-Butyne: (Dimethylacetylene)	54.				-32	27								171	454
Butyraldehyde:(butanal)	72.11	2 3 1		0 0 0	-99	75.7	88.3	5.3	7.0					154	437
n-Butyric acid:	88.11	2 2 0		1 3 1	-4.5	165.5	92.0	4.1	10.6	4.8				126	477
Butyric anhydride:	158.2	1 2 1		0 0 0	-53.5	181.5	165.9	10.2	8.9						316
Butyrolactone: (gamma)	86.09	0 1 0		1 1 1	-42	206	76.3	16.6	7.4						608
Butyronitrile:	69.11	3 3 0			-112	118	87.1	12.5	5.1					73	481
C*															
Calcium carbonate	100.09			1 0 1	825d						8.3				
Calcium chloride:	110.99				782	>1600									
Calcium fluoride	78.08			1 1 0	1330	..					10.3				
Clacium formate:	130				d										
Calcium nitrate:	164				561										
Calcium oxalate	128.1			1 3 1	d						8.6				
Calcium sulfate	136.14			1 1 0	1450						5.9				
Camphor:	152.2				179	207								45	392
Capric acid:	172.26			1 1 2	32	269	194						163	163	
e-Caprolactam:	113.16				69-71	260	110.8					16		142	485
e-Caprolactone:	114.14				-18	215-6	106.5	9.9	14.4						
Carbon dioxide:	44				-78(s)									235	(s)
Carbon disulfide:	76.1	2 3 0	16	3 1 1	-111.5	46.2	60.3	16.6	4.3					58	362
Carbon monoxide:	28	2 4 0	16	0 0 0	-205	-191								30	216
Carbon tetrabromide:	331.6	3 0 1			90	187									
Carbon tetrachloride:	153.8	4 0 ?		3 3 1.	-23	76.5	96.5	0.0	0.6			D		174	195
Carbon tetrafluoride:	88.01				-184	-130									
Carbonyl sulfide:	60.1	3 4 1			-138.2	-50	48.4					79			
2-Carene:	136.24					167									
Cellulose:	(162)x				260										
Cellulose acetate:															
Cellulose triacetate:	288.25														
Cetyl alcohol:	242			1 1 1	49-50	189	295							142	
Chloral:	147.4			? 1 3	-57.5	97.8	97.5								221
Chlorine:	70.9	3 0 1	14	3 3 3	-101	-34	50.6							90	291
Chlorine dioxide:	67.45				-59.5	9.9									
Chloroacetic acid:	94.5				62	189									
Chloroacetyl chloride:	112.94				-22	105									
Chlorobenzene:	112.6	2 3 0	24	3 2 3	-45.6	132	101.8	9.4	0.0			D-A		85	318
1-Chlorobutane:	92.57	2 3 0	16	1 0 3	-123.1	78.4	104.5	6.9	3.5						325
2-Chlorobutane:	92.57	2 3 0			-131.3	68.2	106	6.9	2.4						317
Chloro-m-cresol:	142.6				48	225									

continued

Table Da-1: Selected Properties for Selecting Separation Processes (at 25°C unless noted)

	mol mass	NFPA H F S	Do	Cuc s s s	freezing temp. °C	boiling temp C	Molar volume	δ_p $(J/cm^3)^{0.5}$	δ_h	pKa	pKsp	Biodegr mg COD h.g	BOD/ COD	ΔH_m kJ/kg	ΔH_v kJ/kg
Chlorodifluoromethane:	86.47		1	0 0 0	-146	-40.8	72.9	6.3	5.7						
2-Chloroethyl acetate:	122.6	2 2 0				141.5	107.5	11.4	9.6						338
2-Chloroethyl ethyl ether:	108.57					107	109.8	8.5	7.5						
2-Chloroethyl vinyl ether:	106.5				---	109									
Chloroform:	119.38	2 0 0	1	2 2 2	-63.5	61.7	80.5	13.7	6.3			A		80	
Chloromethyl Methyl ether:	80.51				-103	55									
1-Chloronaphthalene:	162.6	1 1 0			-2.3	259									320
2-Chlorophenol:(o-)	128.6	3 2 0			9.3	175									
3-Chlorophenol:(m-)	128.6				33.5	214									
4-Chlorophenol:(p-)	128.6	3 1 0			43	220									
Chloroprene:	88.5	2 3 0		1 1 1	-130	59.4	93.	7.5	3.5						302
3-Chloropropanol:	94.54	2 2 0			---	165	83.6	5.7	14.7						
o-Chlorostyrene:	138.6	2 1 2	24				126.8	9.4	0.0						
p-Chlorostyrene:	138.6						128.3	9.2	0.0						
Chlorosulfonic acid:	116.52				-80	158									
2-Chlorotoluene:(o-)	126.59				-36	157									
3-Chlorotoluene: (m-)	126.59				-48	161									
4-Chlorotoluene:(p-)	126.59				6	162									
Chlorotrifuoromethane:	104.47				-179	-82									
5-Chlorouracil:	146.53				>300										
Citric acid:	192.12			2 3 3	153	d				3.14					
Copper (I) chloride:	99				430	1490					5.9				
Copper (II) hydroxide:	97.59				d						18.8				
Copper (I) sulfide:	159.2				1100										
Copper (II) sulfide:	95.63				103	220 d					35				
Courmarin:	146.15				68.5	298									
Creatinine:	113				255 d										
m-Cresol:	108.14	2 1 0		0 3 0	11.5	202.2	104.6	5.1	12.9	10		55		99	421
o-Cresol:	108.14	2 2 0			30.8	190.8	103					54	0.74	146	418
p-Cresol:	108.14	2 1 0			35	202	104.5					55		110	439
Crotonaldehyde:	70.09	3 3 2		0 1 0	-74	104-5	82.5	10.5	11.5						515
Cumene:	120.19	0 2 0	10	1 1 1	-96	152.4	139.5	6.9	0.0					80	312
Cyanamide:	42.04	4 1 3	29	? 1 1	44-5	140	39							207	
Cyanogen:	52.04				-28	-22									
Cyanogen chloride:	61.48				-7	13									
Cyanogen fluoride:	45.02				?	-73									
Cyclobutane:	56.11	1 4 0	21		-50	12									431
Cyclodecane:	140.27				201	176.5									
Cyclodextrin:	1297				267 d										
Cycloheptane:	98.19			-12	118.5	121									
Cyclohexane:	84.16	1 3 0	16	1 1 1	6.5	80.7	108.1	3.1	0.0					32	356
Cyclohexanol:	100.16	1 2 0	4	0 0 0	25.1	161.1	104.1	4.1	13.5			28		17	455
Cyclohexanone:	98.14	1 2 0		1 1 1	-16.4	155.6	103.5	9.4	11.0			30			411
Cyclohexene:	82.15	1 3 0		1 1 1	-103.5	83	101.2								368
Cyclohexylamine:	99.18	2 3 0			-17	134									367
Cyclohexyl chloride:	118.61	2 3 0			1.0	143	81.1	5.5	2.1						313
1-3-Cyclooctadiene (cis- cis):	108.18				-57		124.5								
1-5-Cyclooctadiene (cis- cis):	108.18				-70	149-50	122.7								334
Cyclopentane:	70.13	1 3 0		1 1 1	-93.9	49.2	94.0	3.9	0.6					9	389
Cyclopentanol:	86.13				-19	139-40	90.7	11.5	16.5						
Cyclopentanone:	84.12	2 3 0			-51.3	130.6	88.7	11.1	8.8						
Cyclopentene:	68.12				-135	44.2	88.2	5.8	4.1						385
Cyclopropane:	42.08	1 4 0			-127.6	-32.7									476
p-Cymene:	134.22	2 2 0			176									72	285
Cytosine:	111				>300 d										

D*

2,4-D:	221.94				136								
DDE:	318												
DDT:	354				109								
cis-Decahydronaphthalene:	138.25	2 2 0			-43	195.6	154.2	0	0			69	297
trans-Decahydronaphthalene:	138.25	0 2 0			-30.4	187.2	158.9	0	0			104	291
Decane: (C10)	142.28	0 2 0			-29.7	174.1	194.9	0.0	0.0			202	276
1-Decanol: (Decyl alcohol)	158.28	0 2 0	5		fr 7	229	190.8	2.7	10.0			238	315
2-Decanol:	158.28				-2.4	211	191.9	6.2	9.8				
1-Decene:	140.27	0 2 0			fr 66.3	170.5	189.3	3.3	1.3			99	276
n-Decylbenzene:	218.4	2 1 0					256.9	5.0	0.0				
Deuterium Oxide:	20.02				3.82	101.4				14.87			
Dextran:	162.14												
Diacetone alcohol:	116.16	1 2 0	5	0 0 0	-44	164	123.7	11.4	12.6				411
Diallylamine:	97.16				-88	111	124.1	8.5	7.5	9.3(+1)			
Diallyl ether: (allyl ether).	98.15	3 3 2	16	1 1 1		94-95	122						
Diallyl phthalate:	246.2												
3-3-Diaminodipropylamine:	131.22				-14		139.9	11.5	12.9				
1,3-Diaminopropane:	74.13	2 3 0			-12	140	83.5	12.9	14.7				
Diamyl phthalate:	306	0 1 0				250(7kPa							
1,2,3,4-Dibenzoanthracene:	278				205	518							
Dibenzyl ether:	198.26	0 1 0		1 1 1	3.6	298	190.1	3.7	7.4			102	
o-Dibromobenzene:	235.91				1.8	225	118.9	10	0.0			53.5	
m-Dibromobenzene:	235.91				-6.9							56	
p-Dibromobenzene:	235.91				86	219						86	
Dibromochloromethane:	208.3				-22	120							
Dibromochloropropane:	113												
Dibromomethane:	173.9				-53	97							
Dibutylamine: (DBA).	129.25	3 2 0			-60	159	168.5	5.6	5.7	11.3(+1)			314
N-N-Dibutylethanolamine:	173.3	3 2 0			-70	222	202.5	6.0	9.9				
Dibutylisopropanolamine:	187.3	2 1 0					223.8	5.8	8.8				
Dibutyl maleate:	228.28	1 1 0	5				230.6	8.6	8.4				260
Di-n-butyl phthalate:	278.35	0 1 0		1 1 1	-35	340	266.9	9.5	8.1		D		285
Dibutyl sebacate:	314.5	0 1 0	5		1	345	339	4.5	4.1				296
Dichloroacetyl chloride:	147.4	3 2 1				107							
o-Dichlorobenzene:	147	2 2 0	29	1 1 1	-17	180.5	112.7	9.8	0.0		T	88	270
m-Dichlorobenzene:	147	2 2 0			-24.7	173	114.1	9.5	0.0		T	86	263
p-Dichlorobenzene:	147	2 2 0			53.1	174	117.8	9.2	0.0		T	124	429
3,3-Dichlorobenzidine:	253												
Dichlorodifluoromethane:	120.91				-158	-29.8		2.0	0.0				
1,1-Dichloroethane:	98.96	2 3 0	16		-97	57.3	84.2	10.5	5.6		A	80	289
1,1-Dichloroethylene:	96.94	2 4 2	24	1 3 0	-122	30-2	79.9	6.8	4.5		A	67	270
cis 1-2-Dichloroethylene:	96.94	2 3 2			-80	60.3						74	312
trans 1-2-Dichloroethylene:	96.94	2 3 2			-50	47.5						124	298
Dichlorofluoromethane:	102.92				-135	9	73.3	3.1	5.7				
Dichloroisopropyl ether:	171.1	2 2 0				187	154.7	9.7	4.6				
Di (2-chloroisopropyl) ether:	171	2 2 0				187	146	8.2	5.1				
Dichloromethane:	84.93	2 0 0	14	2 3 1	-95.1	40	64.0	6.3	6.1		D	54	330
1-Dichloro-2,4-nitrobenzene:	192				61	257							
1,5-Dichloropentane:	141	2 3 0			-72	178							
2,3-Dichlorophenol:	163				59	206							
2,4-Dichlorophenol:	163	- 1 0			45	210							
2,5-Dichlorophenol:	163				59	211							
2,6-Dichlorophenol:	163				68	219							
3,4-Dichlorophenol:	163				66	253							
1,2-Dichloropropane:	113				-110	96							
1-3-Dichloropropane:	113	2 3 0	16		-99.5	120.4							300
1-3-Dichloro-2-propanol:	128.99	2 2 0			-4	174.3	95.5	13.6	16.4				

continued

Table Da-1: Selected Properties for Selecting Separation Processes (at 25°C unless noted)

	mol mass	NFPA H F S	Do	Cuc s s s	freezing temp. °C	boiling temp C	Molar volume	δ_p $(J/cm^3)^{0.5}$	δ_h	pKa	pKsp	Biodegr mg COD h.g	BOD/ COD	ΔH_m kJ/kg	ΔH_v kJ/kg
1-3-Dichloropropene:	111	2 3 0				104									
Dichlorotetrafluoroethane:	170.92			0 0 0	-94	3.5	117	1.8	0.0						
2,4-Dichlorotoluene:	161				-13	201									
Dicyandiamide:	84				207	d									
cis-Dicyano-1-butene:	106														
trans-Dicyano-1-butene:	106														
1,4-Dicyano-1-butene:	106				74										
Dicyclohexylamine:	181.3	3 1 0			-0.1	256									
Dicyclopentadiene:	132				-1	170 d									
Dieldrin	380.9				175-6										
Diesel fuel:		0 2 0	10			157									
Diethanolamine:	105.14	1 1 0	5	3 0 0	28	270	95.7		8.9(+1	8.9(+1)		19.5		239	622
1-1-Diethoxybutane:	146.2					143	177.6	6.2	4.9						
2-5-Diethoxytetrahydrofuran:	160.22					167-74	165.7	7.7	5.9						
Diethylamine:	73.14	3 3 0	16	3 0 1	-50	55	104	7.0	6.3	10.8(+1)				385	
Diethylamine hydrochloride:	109.6				227-30	320-30									
3-(Diethylamino)propylamine:	130.24	2 2 0				157	157.7	7.4	7.1						
N,N-Diethylaniline:	149.24	3 2 0			-21	217									
2,6-Diethylaniline:	149.24				3	243									
1-2-Diethylbenzene:	134.22	2 2 0	10		-31	183	152.5	6.5	0.0						
1-3-Diethylbenzene:	134.22				-84	180-1	155.3	6.3	0.0						
1-4-Diethylbenzene:	134.22				-43	184	155.7	6.2	0.0						
Diethyl carbonate:	118.13	- 3 1	16		-43	126-28	121.2	3.1	6.1						306
Diethyl disulfide:	122.2				-101	154									
Diethylene glycol:	106.12			1 1 0	-10	245	94.9	12.3	23.3	4.4(+3)	13.7	0.14			493
Diethylene glycol-															
-butyl ether:(mono n-)	162.23	1 2 0	5		-68.1	231	169.8	7.0	10.6						307
-butyl ether acetate:	204				-32	246									246
-diacetate:	190.2	0 1 0					171.7	10.7	11.2						
-dibutyl ether:	218.3				-60	256									230
-diethyl ether:	162.23	1 2 0			-40	180-90	178.4	7.1	8.0						
-dimethyl ether:	134				-64	162									309
-ethylbutyl ether:	190.3						204	7.4	11.2						
-ethyl ether:(mono)	134.18	1 1 0		1 1 0	-78	202	134.5	8.9	14.1						354
-ethyl ether acetate:	176.21	1 1 0			-25	217	174.5	9	9.4						278
-ethylhexyl ether:(2-)	218.3						237.8	6.1	7.2						
-hexyl ether:(n-)	190				-40	260	204.8	7.5	12						286
-methyl ether:(mono)	120.15				-85	193	117	9.6	15.8						388
-methyl ether acetate:	162				-99	209	157.5	9.6	10.3						299
-phenyl ether:	182				-40	298	164.3	10.4	14.3						318
-phthalate:	310	0 1 0													
-vinyl ethyl ether:	160				-80	194	171.9	8	8.6						274
Diethylenetriamine:	103.17				-39	207	107.6	13.1	14.7						
N-Diethylethanolamine:	117.19	3 2 0			-70	161	132.6	7.3	12						
Diethyl fumarate:	172.18	1 1 0			1	218	163.7	10.3	9.3						
Di(2-ethylhexyl)amine:	241	3 1 0			35.6	297	267.8	4.2	4.8						
Diethyl ketone:	86.13	1 3 0			-39.8	101.7	105.8	8.7	7.6					135	391
Diethyl maleate:	172.18	1 1 0			-10	225	161.8	10.5	9.9						304
Diethyl oxalate:	146				-40.6	185									
3-3-Diethylpentane:	128.26	0 3 0			-33.1	146.2	170.2	0.0	0.0						
Diethyl phthalate:	222.24	0 1 0			-3	298-99	198.8	10.9	8.5			D			
Diethyl pimelate:	216.28				-24	252	217.6	8.5	8.3						
Diethylsiloxane: (silicone)	400				-70	282									
Diethyl succinate:	174.2	1 1 0			-20	217.7	166.4	9.8	9.8						

Diethyl sulfate:	154.18	2 1 0			-24	208	140	14.7	7.2				
Diethyl sulfide:	90.18				-103.8	92.1	107.8	3.1	2.1			132	353
Difluoromethane:	52				-	-52							
Diglycol chlorohydrin:	124.6	0 1 0				197	106.7	12.1	17.1				
Dihexylamine:	185.35	2 1 0			---	192-95	233.1	4.8	6.1	11(+1)			
Dihexyl ether:	186.33				-43	226	234.8	3.7	5.2				
Diisobutylene:	112.22	1 3 0	16		-93	101-2	158.5	3.4	0.0				
Diisobutyl ketone:	142.24	1 2 0		0 0 0	-42	168	176.6	6.8	3.9				
Diisopropylamine:	101.19	3 3 0	15		-96	84	140.2	6.2	2.0	11.1			326
Diisopropylethanolamine:	145.3	1 2 0				191	167	6.9	10.5				
Diisopropyl maleate:	200.2	1 1 0				229	199.2	9.6	7.2				
Diketene:	84.07	2 2 2				127	78.4	13.5	12.7			150	419
1,1-Dimethoxyethane:	90.12				-113	64	105.8	7.8	6.8				
2,2-Dimethoxypropane:	104				---	83							
N-N-Dimethylacetamide:	87.12				-20	165	93	11.5	10.2			120	494
Dimethyl amine:	45.08	3 4 0	21	3 0 0	-93	7.4				10.8(+1)	3.6		607
4-Dimethylaminoazobenzene:	225				111 d								
3-(Dimethylamino)-													
-propionitrile:	98.15	- 2 1			-43	170	112.8	10.5	11.4				
-propylamine:	102.18	3 2 0			125.8	8. 133	125.8	8.5	7.9				
N-N-Dimethylaniline:	121.18	3 2 0			1.5	193							
2,2-Dimethylbutane:	86.18	1 3 0			-99.9	49.7	132.9	0	0			7	309
2,3-Dimethylbutane:	86.18	1 3 0			-128.5	58	130.3	0	0			9.4	320
1,3-Dimethyl-1-butanol:	102.18	2 2 0		0 0 0	-90	132	127.2	7.6	10.5				
2,2-Dimethylbutanol:	102.18				<-15	136.7	123.4	8.7	12.1				
2,3-Dimethylbutanol:	102.18				---	142	124.2	8.5	12.2				
2,4-Dimethylbutanol:	102.18						126.3	8.5	11.4				
2,3-Dimethyl-2-butanol:	102.18				-14	118.4	124.1	9.4	10.7				
2,3-Dimethyl-1-butene:	84.16	0 3 0			-157	55.7	123.7	3.8	3				
2,3-Dimethyl-2-butene:	84.16	0 3 0			-74	73.2							
3,3-Dimethylbutyl acetate:	144.2	1 2 0				143	167.1	7.1	3				
1,3-Dimethylbutylamine:	101.19	2 3 0				109	141.1	7.3	3.8				
1,1-Dimethylcyclohexane:	112.22					119	144.4	2.6	0				296
cis-1,2-Dimethylcyclohexane:	112.22				-50	129.7							
trans-1,2-Dimethylcyclohexane:	112.22				-88	123.4							
1-1-Dimethylcyclopentane:	98.2				-70	87.5	131.1	3.2	0			14	312
2-6-Dimethyl-1-4-dioxane:	116.2	2 3 0				117	123.7	8.3	4.3				
Dimethyl disulfide:	94.2				-85	110	89.3	13.4	5.4				
N-N-Dimethylethanolamine:	89.14	2 2 0				133	100.5	8.5	14				
Dimethyl ether:	46.07	2 4 0	21	1 1 1	-138.5	-25							460
N,N-Dimethylformamide:	73.1	1 2 0		0 2 0	-61	153	77.4	13.7	11.3	-0.01		221	577
2,5-Dimethylfuran:	96.13	2 3 0			-62	92	106.5	7	4.8				
2,2-Dimethylheptane:	128.26				-113	132.7	180.5	0	0				
2,3-Dimethylheptane:	128.26				-116	140.5	176.7	0	0				
2,4-Dimethylheptane:	128.26				--	133.5	179.6	0	0				
2,5-Dimethylheptane:	128.26				--	136	178.2	0	0				
2,6-Dimethylheptane:	128.26				-102.9	135.2	180.9	0	0				
3,3-Dimethylheptane:	128.26	0 3 0			--	137.3	176.8	0	0				
3,4-Dimethylheptane:	128.26				---	140.1	175.4	0	0				
3,5-Dimethylheptane:	128.26				---	136	177.5	0	0				
4,4-Dimethylheptane:	128.26				---	135.2	177.6	0	0				
2,6-Dimethyl-4-heptanol:	144.26					177	178.3	6.7	9.2				
2,2-Dimethylhexane:	114.23				-121	106.8	164.3	0	0			101	283
2,3-Dimethylhexane:	114.23	0 3 0		1 1 1	---	115.6	160.4	0	0				294
2,4-Dimethylhexane:	114.23	0 3 0			---	111	164.1	0	0				287
2,5-Dimethylhexane:	114.23				-91.2	109	164.7	0	0			113	287
3,3-Dimethylhexane:	114.23				-126.1	112	160.9	0	0			62	287
3,4-Dimethylhexane:	114.23				---	117.7	158.7	0	0				294
1,1-Dimethylhydrazine:	60.1	3 3 1		0 0 0	-58	62	75.9	5.9	11				

continued

Table Da-1: Selected Properties for Selecting Separation Processes (at 25°C unless noted)

	mol mass	NFPA H F S	Do	Cuc s s s	freezing temp. °C	boiling temp C	Molar volume	δ_p $(J/cm^3)^{0.5}$	δ_h	pKa	pKsp	Biodegr mg COD h.g	BOD/ COD	ΔH_m kJ/kg	ΔH_v kJ/kg
N,N-Dimethylisopropanolamine:	103.2	2 3 0				125	122.1	7.7	11.2						
Dimethyl maleate:	144.13	1 1 0				201	125.1	12.1	11.8					102	355
2,6-Dimethylmorpholine:	115.18	2 2 0			-85	147	123.2	10	2.3						
Dimethylnitrosoamine:	74.1	3 - -				154									
2,2-Dimethyloctane:	142.3						197.69	0	0						
2,3-Dimethyloctane:	142.3				---	164.7	192.9	0	0						
2,4-Dimethyloctane:	142.3	0 2 0					197.4	0	0						
2,5-Dimethyloctane:	142.3						194.7	0	0						
2,6-Dimethyloctane:	142.3				---	161	194.6	0	0						
2,7-Dimethyloctane:	142.3				-54.6	159.6	196.6	0	0						
3,3-Dimethyloctane:	142.3						193.9	0	0						
3,4-Dimethyloctane:	142.3					132	192	0	0						
3,5-Dimethyloctane:	142.3						194.7	0	0						
3,6-Dimethyloctane:	142.3						194.7	0	0						
4,4-Dimethyloctane:	142.3						194.4	0	0						
4,5-Dimethyloctane:	142.3						191.8	0	0						
2,3-Dimethylpentaldehyde:	114.2	2 3 0				145	138.8	7.7	6.3						
2,2-Dimethylpentane:	100.2				-123.8	79.2	148.7	0	0					59	292
2,3-Dimethylpentane:	100.2	0 3 0			---	89.8	144.2	0	0						305
2,4-Dimethylpentane:	100.2	0 3 0			-119.2	80.5	149	0	0					67	297
3,3-Dimethylpentane:	100.2				-134.4	86.1	144.5	0	0					71	296
2,3-Dimethyl-3-pentanol:	116.2				<-30	140	131.6	8.2	12						
2,3-Dimethylphenol:	122.2				75	218									
2,4-Dimethylphenol:	122.2				27	210									
2,5-Dimethylphenol:	122.2				74.5	212									
2,6-Dimethylphenol:	122.2				49	203									
3,4-Dimethylphenol:	122.2				62.5	225									
3,5-Dimethylphenol:	122.2				64	220									
Dimethyl o-phthalate:	194.2	0 1 0			2	282	163.2	12.6	9.7						
Dimethyl m-phthalate:	194.2				67	282									
Dimethylpimelate:	188.22				-21		180.8	9.4	9.6						
2,2-Dimethylpropane:	72.2	0 4 0			-16.6	9.45	123.6	0	0					45	316
2,2-Dimethyl-1-propanol:	88.15	2 3 0	16		52	113	107.6								
Dimethyl sulfide:	62.13	4 4 0			-83	38	73.4							129	452
Dimethyl sulfone:	94.13				110	238	75	19.4	12.3						
Dimethylsulfoxide:	78.13	1 1 0			18.4	189	70.9	16.4	10.2					179	553
Dimethyl terephthalate:	194.2				140	-									
2,5-Dimethyltetrahydrofuran:	100.16				-	91	120.2	6.5	0						
o-1-2 Dinitrobenzene:	168.1	3 1 4			116.9									135	
m- 1-3 Dinitrobenzene:	168.1				89.7	297	122							103	
p- 1-4 Dinitrobenzene:	168.1				173.5		103							167	
4,6-Dinitro-o-cresol:	198.2				87.5	-									
2,4-Dinitrophenol:	184				112										
2-4 Dinitrotoluene:	184.14	3 1 4			70.14	300 d								110	
Dioctyl phthalate:	390.56	0 1 0		0 0 0	-50	384	398.1	7	3.1						
Dioctyl sulfosuccinate-sodium salt:	444.5				173										
1-3 Dioxane:	88.1				-45	105									
1,3-Dioxolane	74.1				-95	74									
Dipentene:	136.24	0 2 0		0 0 0		178	162.2	5.8	0						
Diphenylamine:	169.23	3 1 0			53	302									
Diphenyl ether:	170.21	1 1 0	14	0 0 0		258	158.4	11.2	0					101	277
1,2-Diphenyl hydrazine:	184.2				123										
Diphenylmethane: (Ditane)	168.24	1 1 0			25.3	264.3	167.2	9.4	0					111	
Dipropylamine:	101.19	3 3 0			-40	110	136.7	6.2	5.8	11(+1)					345

Dipropylene glycol:	134.18	0 1 0	4			230	131.2	10.3	17.4				400
(mixed isomers)...							131.3	20.3	18.4				
-butyl ether:(mono n-)	190.3					229							
-ethyl ether:(mono)	162.2					338							
-isopropyl ether:	176.2					80							
-methyl ether:(mono)	148.2	0 2 0			-117	188.3							
Dipropylether:	102.18				-123	91	138.8						324
Dipropyl ketone:	114.19	2 2 0			-33	144	139.7	7.5	6.9				317
Dipropyl oxalate:	174.2				-44	211							
Dipropyl sulfone:	150.2				28	270							
Divinyl acetylene:	78.1	1 3 3	29			85	85						
Divinyl benzene:	130.2	1 2 2	24			195	142						
Divinyl ether:	70	2 3 2				39							
Docosanoic acid:	340.59				79.9							231	
Dodecane: (Dihexyl)	170.34	0 2 0			-9.6	216.3	227.5	0	0			215	257
Dodecanoic acid:	200			? 0 ?	43	298						183	289
1-Dodecanol:	186.34	0 1 0		0 0 0	25	261	227.2	6.5	10.8				
2-Dodecanol:	186.34				19	252	224.9	5.8	9.2				
1-Dodecene:	168.32	0 1 0			-35	213							
Dodecylamine:	185				28	247							
Dodecyl benzene:	246.5				-7	331							
Dowtherm A:	166	1 1 1	14		12	257	157.2				0.67	98	296
Dowtherm J:	134				-73	181	154						304
E*													
n-Eicosane: (Didecyl)	282.55	- 1 0			36.8	343	358.3	0	0			218	
Eicosanoic acid:	312.5				75.3							251	
a-Endosulfan:	404												
Endrin:	378												
Epichlorohydrin:	92.53	3 2 1	24	3 3 0	-57	116	78.2	12.3	10.5			112	412
3,4-Epoxy-1-butene:	70.09					66	80.6	9.8	8.6				
Ethane:	30	1 4 0	21	0 0 0	-183	-88	93					95	489
Ethanol: (Ethyl alcohol)	46.07	0 3 0	16	0 0 0	-117.3	78.5	58.4	11.2	20	15.9	0.75	108	848
2-Ethoxy-3-4-dihydro-2-pyran:	128.2	2 2 1				143	133.1	8.8	4.9				
3-Ethoxy-1-propanol:	104					160-61	115.2	8.7	14.8				
3-Ethoxypropionaldehyde:	102	2 3 0				135	112.1	9.7	10.5				
3-Ethoxypropionic acid:	118	2 1 0				219	113.4	11.5	16.5				
Ethyl acetate:	88.1	0 3 0	16	0 1 0	-84	77	90.4	8.6	8.9		0.81	119	360
Ethyl acetoacetate: enol form	130.14			0 0 0	-43	181	127.5	11.1	10.2	10.68			
Ethyl acrylate:	100.12	2 3 2	16	0 0 0	-71	99	108.4	9.3	7.9				343
Ethyl amine:	45.08	3 4 0		3 3 1	-81	16	66			10.6(+1)	0.6		607
Ethyl amyl ketone:	128.22					159	155.8	7.2	5.7				
N-Ethylaniline:	121.18	3 2 0			-63	205	125.8	10.5	7.7	5.1(+1)			335
Ethylbenzene:	106.17	2 3 0	16	1 2 1	-95	136.2	122.5	0.6	1.4			86	343
Ethyl benzoate:	150.78	1 1 0 4	4	0 0 0	68	213	144.2	10.6	6.3				270
Ethyl bromide: (Bromomethane)	108.97	2 3 0	21	0 0 0	-118.6	38.4	74.6	5.1	6.6			54	243
2-Ethyl-1-3-butadiene:	82.2						115.3	5.6	0				
2-Ethyl-1-butanol:	102.18	1 2 0			-15	146	123.1	8.5	13				423
2-Ethyl-1-butene:	84.16	0 3 0			-131.5	64	122.1	3.9	3.1				
2-Ethylbutyl acetate:	144.2	1 2 0					165	7	5.3				
N-Ethylbutylamine:	101.19	3 3 0	16		20	125	130.4	6.2	5.9				
Ethylbutyl ether:	102.17	2 3 0				65	135.8	4.8	4.5				
Ethylbutyl ketone:	114.19	1 2 0			-39	147	139.5	7.4	6.9				
2-Ethylbutyraldehyde:	100.2	2 3 1					123.7	7.9	7.2				
Ethyl butyrate:	116.16	0 3 0 1	16	0 0 0	-93.3	120	132.3	7.6	6.4				313
2-Ethylbutyric acid:	116.16	2 1 0			-14		125.7	9.3	12.9				
Ethyl chloride:	64.5	2 4 0	21	0 3 3	-139	13	70.3					69	385
Ethyl chloroacetate:	122.55	2 2 0			-26	143							

continued

Table Da-1: Selected Properties for Selecting Separation Processes (at 25°C unless noted)

	mol mass	NFPA H F S	Do	Cuc s s s	freezing temp. °C	boiling temp C	Molar volume	δ_p $(J/cm^3)^{0.5}$	δ_h	pKa	pKsp	Biodegr mg COD h.g	BOD/ COD	ΔH_m kJ/kg	ΔH_v kJ/kg
Ethyl chloroformate:	108.52	- 3 1			-81	93	95.6	10	6.8						
Ethyl cinnamate:	176.22				6.5-7.5	271	168	8.2	4.1						
Ethyl crotonate:	114.1	2 3 0				142	124.3	8.8	7.6						
Ethylcyclohexane:	112.22	1 3 0			-111	130-32	142.4	2.6	0					74	308
N-Ethylcyclohexylamine:	127.23	3 3 0				165	150.7	7	3						
Ethylcyclopentane:	98.19	1 3 0			-138	103	128.7	3.3	0					46	328
N-Ethyldiethanolamine:	133.19	2 1 0			-50	246-52	131.4	9.5	18.2						
Ethylene: (Ethene)	28	1 4 2	24	0 0 0	-169	-103								120	483
Ethylene carbonate:	88.06	2 1 1	14		35-37	243-4	66.7	21.7	5.1					114	570
Ethylene chlorohydrin:	80.51	3 2 0		1 1 1	-89	129	67	13.3	19.6	14.3					518
Ethylene cyanohydrin:	71.08	2 1 1		1 3 1	-46	228	68.3	15.2	24.7						790
Ethylenediamine:	60.10	3 2 0		3 1 0	8.5	116.5	66.8	8.8	17	4		9.8		322	698
Ethylenediamine tetra- acetic acid:	292				252 d										
Ethylene dibromide:	187.86			3 3 1	9.8	131.3	86.2	6.8	12.1					57	194
Ethylene dichloride:	99.96	2 3 0	15	3 3 1	-35.3	83.5	80.1	11.2	9.1			B		89	324
Ethylene glycol:	62.07	1 1 0		1 1 0	-11.5	198	56	15.1	29.8	14.2		41.7	0.44	188	800
-benzyl ether:	152.2				<-75	256	143	10.2	13.8						
-butyl ether:(mono n-)	118.18	0 2 -		1 1 0	-70	170	132.8	5.1	12.3						367
-butyl ether acetate:	160.2				-63	192									274
-butyl methyl ether:	132.2						157.2	6.3	6.7						
-diacetate:	146.14				-41	186-7									311
-dibutyl ether:	174.3				-69	203	209.5	6.1	16.8						253
-diethyl ether:	118.2				-74	121	144	6.8	5.1						305
-diformate:	118.09				-10	177									374
-dimethyl ether:(1,2-)	90.12	- 0 2			-71	85	104.5	7.7	7.7						349
-dipropionate:	174.2					210									269
-ethylbutyl ether:(2-)	146.2				-90	196.7	164.3	7.3	11.5						
-ethyl ether:(mono)	90.12	2 1 0			-90	135	96.9	9.2	14.3						450
-ethyl ether acetate:	132.2	2 2 0			-61	156	135.5	9.0	8.9						309
-ethyl ether acrylate:	144.2				-47.3	174	147.7	9.4	8.7						295
-ethylene ether:	88.11	2 3 0	16	1 0 0	11.8	101	85.2	10.1	7					146	401
-ethylhexyl ether:(2-)	174.28				-70	229	198.1	6.7	10.6						
-hexyl ether:(n-)	146.23				-50	208	165.6	7.2	12.1						328
-methylene ether:	74.08	2 3 2	24		-95	78	69.9	11.3	13.9						
-methyl ether:(mono)	76.1	2 2 0		1 1 0	-65.1	125	78.9	9.2	16.4	14.8					519
-methyl ether acetate:	118.1	0 2 0			-65	145	118.2	9.8	9						340
-methyl ether acrylate:	130.14				-44.6	156									
-methyl ethyl ether:	104					102									
-phenyl ether:	138				14	244.9	125.3	10.4	16						375
-phenyl ether acetate:	180				-2.7	259.2	163.7	11.1	8.8						291
-vinyl butyl ether:	144.2				-71 d										
Ethylene imine:	43.07	3 3 3			-74	57				8.04					770
Ethylene oxide:	44	2 4 0	29	0 0 1	-112	10.5	51								580
N-Ethylethanolamine:	89.1	1 2 0				161	97.6	10.5	17.8						
Ethyl ether:	74.12	2 4 0	21		-116.2	34.5	103.8	5.3	5.6					99	351
Ethyl-3-ethoxyacrylate:	144.17					195-96	144.5	8.5	8						
Ethyl formate:	74.08	2 3 0	16	1 3 1	-80	52-54	80.8	7.2	7.6					124	443
3-Ethylheptane:	128.3						177.7	0	0						
4-Ethylheptane:	128.3						177	0	0						
2-Ethylhexaldehyde:	128.2						157.3	7.1	6.5						
3-Ethylhexane:	114.2					119	161.3	0	0						300
2-Ethylhexanoic acid:	144.21					228	159.7	8.3	11.6						
2-Ethyl-1-hexanol:	130.23	2 2 0			-76	183-6	156.3	3.3	11.9						348
2-Ethyl-1-hexene:	112.2						155.4	3.6	0						

2-Ethylhexyl acetate:	172.3	2 2 0					198.7	6.3	5.4					253
2-Ethylhexyl acrylate:	184.28	2 2 1				215-9	208.2	6.8	4.5					
2-Ethylhexylamine:	129.25	2 2 0			-76	169	163.8	6.6	5.9					
2-Ethylhexyl chloride:	148.7	2 2 0				173	169.6	5.4	2.4					
Ethyl iodide:	156				-111	72.3								191
n-Ethyl lactate:	118.13	2 2 0			-26	154	113.4	7.6	12.5					394
Ethyl mercaptan:	62.13	2 4 0	21	3 3 1	-144.4	35	74	6.6	7.2		10.6			460
Ethyl methacrylate:	114.15					118								
3-Ethyl-2-methylhexane:	128.3					138								
3-Ethyl-3-methylhexane:	128.3					140.6								
4-Ethyl-2-methylhexane:	128.3					133.8								
4-Ethyl-3-methylhexane:	128.3													
3-Ethyl-2-methylpentane:	114.2				-115	115.6								
3-Ethyl-3-methylpentane:	114.2				-91	118.2								
N-Ethylmorpholine:	115.18	2 3 1			-63	139	127.3	7.6	0.7					321
1-Ethylnaphthalene:	156.2				-15	258-60								
2-Ethylnaphthalene	156.2				-70	251-52								
3-Ethyloctane:	142.3	0 2 0				167.2	193.6	0	0					
4-Ethyloctane:	142.3	0 2 0				164.4	193.6	0	0					
Ethyl oxalate:	146				-40.6	185								288
3-Ethylpentane:	100.2				-118.6	93.5	144.6	0	0				94	311
Ethyl propionate:	102.13	- 3 0			-73	99	114.6	8.1	7.8					336
2-Ethyl-3-propylacrolein:	126.2	2 2 1				175	149.2	8.4	6.6					
2-Ethyl-3-propylacrylic acid:	142.2	2 1 1				232	150.8	10.1	12.9					
Ethyl propyl ketone:	100.16					123	122.9	7.9	7.3					
Ethyl salicylate:	166.18				2	231								
m-Ethylstyrene:	132.2				-101	190	148.7	7.7	0					
Ethyl sulfide:	90.1				-103	92							132	352
o-Ethyltoluene:	120.2	- 2 0				165	137.4	6.9	0					
m-Ethyltoluene:	120.2	- 2 0				161	139.9	6.7	1.5					
p-Ethyltoluene:	120.2	- 2 0				162	139.9	6.6	2.1					
Ethyl valerate:	130.2				-91.2	145.5								
Eugenol:	164.2				-9.2	255								
F*														
Fluoranthene:	202.26				107-110	384								
Fluorene:	166.22				114.8	295								
Fluorine:	38	4 0 3	29	3 3 3 -2	-188								172	
Fluorobenzene:	96.1	2 3 3	24		-42	85	93.8	9.0	5.2				118	325
Fluoromethane:	34.04				-141.8	-78.4								
5-Fluorouracil:	130.08				280									
Formaldehyde:	30.03	2 4 0	24	1 3 0	-92	-21	36.8							779
Formamide:	45.04			1 1 1	2.5		39.7	26.2	19.0	-0.5			149	
Formamidoxime:	60				112									
Formic acid:	46.03	3 2 0	4	2 3 2	8.4	100.7	37.7	11.9	16.6	3.7			247	502
Fuel T-5:		0 2 0	10											
Fuel oil: No. 4 (API 23):		0 2 0	10											
No. 6 (API 12-16):		0 2 0	10											
Fumaric acid:	116				286					3				
Furan: (Furfuran).	68.08	1 4 1	21			31-32	71.6	1.8	5.3					402
Furfural:	96.09	1 2 1		3 3 1	-38.7	161.7	82.9	14.9	5.1		37		150	449
Furfuryl alcohol:	98.10	1 2 1		0 0 0	-29	171	86.8	7.6	15.1				134	548
G*														
Gasoline:		1 3 0	16											
Glucose:	180.16				146	...								
Glycerol:	92.09	1 1 0	4	2 3 0	20	290	73.0	12.1	29.3	14.4		85	199	663

continued

Table Da-1: Selected Properties for Selecting Separation Processes (at 25°C unless noted)

	mol mass	NFPA H F S	Do	Cuc s s s	freezing temp. °C	boiling temp C	Molar volume	δ_p $(J/cm^3)^{0.5}$	δ_h	pKa	pKsp	Biodegr mg COD h.g	BOD/ COD	ΔH_m kJ/kg	ΔH_v kJ/kg
Glyceryl triacetate:	218.21	1 1 0			-78	258-60	188.2	11.6	11.2						
Guanine:	151				250 d										
H*															
Helium:	4				-272.2	-268.9									
Heptachlor:	370				95										
Heptadecane:	240.47				22	301.8	309							168	220
Heptadecanoic acid:	270.46				61.3	363.8	317.07							190	
n-Heptane:	146.5	1 3 0	16	0 0 0	-91	98	146.5	0.0	0.0					140	320
2-4-Heptanediol:	132.2						142.8	9.9	16.8						
Heptanoic acid:	130.19				-10.5	223	141							115	
1-Heptanol:	116.2				-34.1	176	141.4	8.0	13.0					114	414
2-Heptanol:	116.2					160	142.3	7.4	11.7						429
3-Heptanol:	116.2				-70	157	141.2	7.5	11.8						
2-Heptanone:	114.19	1 2 0			-35	151	139	7.6	7.2						346
1-Heptene:	98.19				-119	94	140.9	3.6	2.7						
cis-2-Heptene:	98.19					98.5	138.7	3.4	2.3						
trans-2-Heptene:	98.19				-109.5	98	140.0	3.1	2.9						
cis-3-Heptene:	98.19					95.8	139.0	6.8	2.5						
trans-3-Heptene:	98.19				-136.6	95.7	140.7	3.1	2.3						
n-Heptylbenzene:	176.29				-37	265	205.4	5.6	0.0						
Hexachlorobenzene:	284.78				231	323									
Hexachlorobutadiene:	260.76	2 1 1	14		-19	210	158								
1-2-3-4-5-6-Hexachloro-															
cyclohexane: (a-isomer)	290.83					156									
(g-isomer- Lindane)	290.83					288									
Hexachlorocyclopentadiene:	272.77				-10	239	160								
Hexachloroethane:	236.74				187										
Hexacosanoic acid:	396.7				87.7										
n-Hexadecane: (Cetane)	226.45	0 1 0			18.2	287	292.8	0	0						
Hexadecanoic acid:	256.43				63.1	351.5	300							164	247
1-Hexadecanol:	242				49.3	344									
1,5-Hexadiene:	82.15				-140.7	59.5	119								
2-4-Hexadienol:	98.15				31-33		112.7	10.4	11.2						
Hexaldehyde:	100.16	2 3 1			-56	128	123.1	7.9	16.9						
Hexamethylbenzene:	162.28				165	263									
Hexamthylene diamine:	116.2				41.2	200	136.8								
Hexamethyleneimine:	99.8				?	138									
Hexamthylenetetramine:	140.19			3 3 0	285-95	s	105.3								
Hexamethyl phosphoramide:	179.2				7	230-32	174.0	8.6	11.2					95	316
n-Hexane:	86.18	1 3 0	16	0 0 0	-95	69	130.5	0.0	0.0					151	337
2,5-Hexanediol:	118.18	2 1 0				216-18	123.0	10.7	19.3						
2-5-Hexanedione:	114.15	1 1 0			-6--5	191	117.3	11.6	11.6						
Hexanoic acid:	116.16	2 1 0			-3	202-03	125.3	9.2	13.8	4.8				129	484
2-Hexanone:	100.16	2 3 0			-57	127	123	7.9	7.1						
1-Hexene:	84.16	1 3 0			-139.8	63.3	125.0	3.9	0.0						
cis-2-Hexene:	84.16	0 3 0			-141.3	68.8	122.5	3.5	3.1						
trans-2-Hexene:	84.16				-133	68-69	124.1	3.3	2.9						
cis-3-Hexene:	84.16				-137.8	66.4	123.8	3.5	3.4						
trans-3-Hexene:	84.16				-113.4	67.1	124.3	3.3	3.2						
Hexyl acetate:	144.21	1 2 0			-80	168-70	164.6	7	6.4						
n-Hexyl alcohol:	102.18	1 2 0		0 0 0	-46.7	158	125.6	8.5	13.7					151	483
2-Hexyl alcohol:	102.18	0 2 0				140	125.2	7.9	12.3						
3-Hexyl alcohol:	102.18					135	124.9	7.7	11.6						

Hexylamine:	101.19	2 3 0			-19	130	132.1	7.1	6.9	10.6(+1)				
Hexylbenzene:	162.26				-67	263								
Hexyl chloride:	120.62	- 3 0			-94	134.5	137.3	6	4					
Hexylcyclohexane:	168				-40	228								
Hexylcyclopentane:	154				-79	203								
Hexylene glycol:	118.18	1 1 0			-40	197	127.8	11.4	16.8					
Histidine:	155.16				285d									
Homopiperazine:	100.17				38	169								
Hydrazine:	32.05	3 3 2	40	3 3 0	41.4	113.5	31.7			5.89				1256
Hydrogen:	2.016	0 4 0	21	0 3 0	-259.1	-252.7	28.4							769
Hydrogen bromide:	80.92			? 3 ?	-86.9	-66.8	46.8			-9				217
Hydrobromic acid:48%														
Hydrogen chloride:	36.47			3 1 1										
-hydrochloric acid,30%wt:	36.47			3 3 3	-111	-85	28.7			-6.1				443
-hydrochloric acid,10%wt:														
Hydrogen cyanide:	27.03	4 4 2	29	3 0 3	-13.2	25.7							311	954
Hydrogen fluoride:	20.01				-35	19.4	20.2			3.18				334
Hydrofluoric acid: 35%														
Hydrogen iodide:	127.9			3 3 3	-50.8	-35.5				-9.5				155
Hydrogen peroxide:	34.02	2 0 2	24	3 3 1	-0.89	151.4	23.66			11.65				1347
Hydrogen sulfide:	34.08	3 4 0	21	3 3 0	-82.9	-59.6	28.65			6.97				
Hydroquinone:	110.11			3 1 1	172	285	82.6				54	0.53	246	
N-(2-Hydroxyethyl)- -morpholine:	131.18						121.1	8.7	13.8					
N-(2-Hydroxyethyl)- -piperidine:	129.2				46-47		122.3	11.0	15.3					
Hydroxylamine:	33.03	1 3 3	29		34	56(3kPa)	24.4			7.99				
I*														
Indane:	118.18				-51.4	178	122.6							
Indene: (Indonaphthalene)	116.16				-1.8	182.6	116.6							
Iodine:	253.84				113.5	184.35	51.5							
Iodobenzene:	204.02				-32	188	111.9						48	194
Iodoform:	393.73				120-23									
1-Iodopropane:	170				-101	102								
2-Iodopropane:	170				-90	90								
Isoamyl acetate:	130.19	1 3 0		0 0 0	-78	142	148.6	3.1	7.0					289
Isoamyl alcohol:	88.15	1 2 0			-117	130	109.0	9.3	14.1					501
sec-Isoamyl alcohol:	88.15				113-4	108.4	8.4	12.4						475
Isobutane:	58.12	1 4 0	21		-159.4	-11.63	105.9	0	0				78	367
Isobutanol:	74.12	1 3 0	16		-108	108	92.3	9.8	15.0				85	569
Isobutyl acetate:	116.16	1 3 0		0 0 0	-98.9	115-17	133.8	7.8	5.1					309
Isobutyl acrylate:	128.2						145.0	8.4	5.7					
Isobutylamine:	73.14	2 3 0	16		-85	68-9	99.7							421
Isobutyl benzene:	134.22	2 2 0			-51	170	157.4	6.3	0.0					
Isobutylene:	56	1 4 0			-140	-6.9	94	4.2	3				106	394
Isobutylene oxide:	72.1	1 1 1				52	90	8.4	1.3					
Isobutyl isobutyrate:	144.21				-81	147-48	168.7	7.0	0.0					265
Isobutyraldehyde:	72.11	2 3 1			-65	63	90.8	9.0	8.4					
Isobutyric acid:	88.11	1 2 0			-47	153-54	92.7	9.9	16.2				57	504
Isobutyronitrile:	69.11				-72	107-08	90.9	10.6	10.6				470	545
Isodecyl alcohol:	158.3	0 1 0					189.6	6.9	10.7					
Isooctyl alcohol:	130.23						157.5	7.3	12.9					
Isopentane:	72.15	1 4 0			-159	30	116.4	0	0				71	339
Isophorone:	138.21	2 1 0		1 0 0	-80	213-14	149.7	9.4	3.2					314
Isoprene:	68.1	2 4 1	21		-146	34	100	5.8	4					383
Isopropanolamine:	75.11	2 2 0			-2	160	96.11	12.3	20.1				164	615
Isopropyl acetate:	102.13	1 3 0	16	0 0 0	-73	88.2	117.1	8.4	5.7					324

continued

Table Da-1: Selected Properties for Selecting Separation Processes (at 25°C unless noted)

	mol mass	NFPA H F S	Do	Cuc s s s	freezing temp. °C	boiling temp C	Molar volume	δ_p (J/cm^3)$^{0.5}$	δ_h	pKa	pKsp	Biodegr mg COD h.g	BOD/ COD	ΔH_m kJ/kg	ΔH_v kJ/kg
Isopropyl alcohol:	60.10	1 3 0		0 1 0	-89.5	82.4	76.6	15.8	6.0			52	0.08	88	667
Isopropylamine:	59.11	3 4 0	21		-101	33-34	85.2	8.6	7.0	10.6(+1)					460
Isopropylbenzoate:	164.2					218									
Isopropyl chloride:	78.54	2 3 1	21	0 0 0	-123	46-47	88.0	7.3	4.3						
Isopropylcyclohexane:	126.24					155	157.4	2.5	0.0						
Isopropylcyclopentane:	112.2				-111	126									
Isopropyl ether:	102.18	2 3 1	16		-85	68-69	140.9	4.7	1.5					108	287
Isovaleraldehyde:	86.13				-51	92	108.3	8.5	7.5						
Isovaleric acid:	102.13				-29	175									
K*															
Kerosene: (No. 1 Fuel Oil)		0 2 0				151-300									
Ketene:	42.04	(2 ? 2)			-151	-56								169	456
Krypton:	83.8				-158	-154									
L*															
Lactic acid:	90.08			3 3 1	16.8	122				3.08					
Lactose:	342				253										
Linoleic acid:	280.44				-9.5										
Lysine:	146.19					224 d									
M*															
Maleic acid:	116.7	3 1 1		2 3 1	130.5	135d	60.7			1.9			0.8		
Maleic anhydride:	98.06	3 1 1	14	1 1 0	57-60	202	65.3							139	559
Malic acid:	134.09				100										
Mercuric chloride:	271.5				277	304									
Mercuric oxide:	216.6				500 d										
Mercury:	200				234	630									
Mesitylene:	120.2	0 2 0			-45	162-64	139.1	7.0	0.6					79	325
Mesityl oxide:	98.15	3 3 0		0 0 0	-53	128	114.4	9.6	6.6						329
Metaldehyde:	176	1 3 1		0 0 0	190-6	s									
Methacrylaldehyde:	70.09	3 3 2	24		-81	69	82.8	10.7	8.8						
Methacrylic acid :	86.09	3 2 2	24		16	163	84.8	13.0	16.2						
Methallyl cyanide:	81.1						97.8	11.5	10.5						
Methane: (Marsh gas).	16.04	1 4 0	21	0 0 0	-182.6	-161.5								59	510
Methanol:	32.04	1 3 0	16	1 1 0	-93.9	65	40.5	13	24	15.5			1.07	99	1088
Methionine:	149.2				273 d										
1-Methoxy-1-3-butadiene:	84.12					91	101.3	8.0	8.2						
3-Methoxybutanol:	104.15	1 2 0				158	112.2	8.8	14.3						
3-Methoxybutyl acetate:	146.2	1 2 0				135-72	154.1	8.6	7.5						
3-Methoxybutraldehyde:	102.1	0 2 0				128	110.9	9.9	9.9						
3-Methoxypropionitrile:	85.1	4 2 1				164-5	91.1	12	14.5						
Methyl acetate:	74.08	1 3 0	16	0 0 0	-98	57.5	79.5	9.5	10.4						410
Methyl acetoactate:	116.12	2 2 0		0 0 0	-80	169-70	107.9	12.2	12.0						
Methyl acetylene:	40	2 4 2	40		-101.5	-23.2	56.6								
Methyl acrylate:	86.09	2 3 2	24		-75	80	90.1	10.2	9.4						335
Methylal:	76.10	2 3 2		0 0 0	-105	41-42	88.5	1.8	8.6						
Methylamine:	31.06	3 4 0	21	3 1 1	-93.5	-6.3	46.9			10.6(+1)				197	820
N-Methylaniline:	107.2				-57	196									
2-Methylaziridine:	57.1					66									
Methyl benzoate:	136.15	0 2 0		0 0 0	-12	198-99	123.5	11.5	7.2						
a-Methylbenzylamine:	121.18	2 2 0				185	128.9	11.1	5.8						

a-Methylbenzyldimethylamine:	149.2	2 2 0					166	7.3	0				
Methyl borate:	103.9	2 3 1	16		-34	68-9	113.5						
Methyl bromide:	94.9				-93.4	3.7						63	
3-Methyl-1-2-Butadiene:	68.1				-113.6	40.9	98						
2-Methyl-1-butene:	70.14	2 4 0			-137	31	107.9	4.1	3.7			113	
2-Methyl-2-butene:	70.14	2 3 0			-134	35-38	106.0	4.1	3.8			108	
3-Methyl-1-butene:	70.14	2 4 0			-168.5	20	111.9	3.8	3.7			75	
N-Methylbutylamine:	87.17	3 3 0				90.5-1.5	118.4	6.7	6.6				
Methyl tert-butyl ether:	88				-111	58							314
Methyl-sec-butyl ketone:	100.2					118	124.0	8.0	6.2				
Methyl tert-butyl ketone:	100.2				-52	106	125.1	8.2	4.1				
Methyl butyrate:	102.13				-85	102							
Methyl carbonate:	90.08	2 3 1	16		0.5	89-90	84						369
Methyl chloride:	50.49	2 4 0	21	0 3 0	-97.1	-24.2	55.0	6.1	3.9			130	428
Methyl chloroacetate:	108.53	2 2 1	14		-32.7	130	87.2						
Methylcyclohexane:	98.19	2 3 0	16	1 1 1	-126	101	127.5	0	1.0			69	322
1-Methylcyclohexanol:	114.19	0 2 0			26	156							308
cis-2-Methylcyclohexanol:	114.19	0 2 0			6.8	165							426
trans-2-Methylcyclohexanol:	114.19				-4.3--3.7	166.5							466
cis-3-Methylcyclohexanol:	114.19	0 2 0			-6--5	168							
trans-3-Methylcyclohexanol:	114.19				-1	167							347
Methylcyclopentadiene:	80	1 2 1	14			73							
Methylcyclopentadiene- dimer:	160				-51	200	170						
Methylcyclopentane:	84.16	2 3 0			-142	72	112.4	3.5	0.0			82	348
N-Methyldibutylamine:	143.3						189.8	3.2	2.9				
N-Methyldiethanolamine:	119.16	1 1 0				246-48	114.8	11.0	20.1				
Methylene diiodide:	267.84				6.1	182	80.5	3.9	5.5				
N-Methylethanolamine:	75.11	2 2 0				159	80.3	11.4	20.0				
Methylethyl ether:	60.1	2 4 0					86.8	5.7	6.2				
Methyl ethyl ketone:	72.11	1 3 0	16	1 1 1	-86.3	79.6	98.5	9.3	8.5		0.82	117	431
2-Methyl-5-ethylpyridine:	121.2	3 2 0			118.5	159	132.3	7.7	2.5				
N-Methylformamide:	59				-3.8	180				-0.04			
Methyl formate:	60.05	2 4 1	21	1 1 1	-99	32	61.6					124	470
2-Methylheptane:	114.23				-109	117.6	163.7	0.0	0.0			90	294
3-Methylheptane:	114.23				-120.5	119	161.8	0.0	0.0			100	299
4-Methylheptane:	114.23				-121	117.7	162.1	0.0	0.0			95	297
2-Methyl-1-heptene:	112.22				-90.1	118.2	155.7	3.6	0.0				
2-Methyl-2-heptene:	112.22					122.6	155.0	3.5	0.0				
3-Methyl-1-heptene:	112.22						159.0	3.4	1.3				
3-Methyl-2-heptene:	112.22						155.0	3.5	0.0				
4-Methyl-1-heptene:	112.22						157.6	3.5	0.0				
5-Methyl-1-heptene:	112.22						157.8	3.4	2.0				
cis-5-Methyl-2-heptene:	112.22						156.3	3.2	0.0				
6-Methyl-1-heptene:	112.22						158.7	3.4	2.2				
cis-6-Methyl-2-heptene:	112.22						157.4	3.2	0.0				
2-Methylhexane:	100.21				-118	90	147.6	0.0	0.0			92	306
3-Methylhexane:	100.21				-119	91	145.9	0.0	0.0				307
5-Methyl-2-hexyl alcohol:	116.2					148-50	141.9	7.6	11.3				
Methyl hexyl ketone:	128.21				-16	173	156.3	7.2	6.8				309
Methyl hydrazine:	46.07	3 3 1	24			87							
Methyl 3-hydroxybutyrate:	118.13	1 2 0					112.0	10.9	14.6				
Methyl iodide:	141.9				-66.5	42.4							192
Methyl isoamyl ketone:	114.19	1 2 0				144	128.6	7.7	5.8				
Methyl isobutyl ketone:	100.16	2 3 0	16	1 1 1	-84.7	116.8	125.5	8.1	5.9				343
Methyl isocyanate:	57.05				-17	37-39	59.0	12.2	13.6				
Methyl isopropenyl ketone:	84.1	2 - 0				98	99.3	10.3	7.1				
Methyl isopropyl ketone:	86.13				-92	94-95	107.0	8.8	6.6				376
Methyl mercaptan:	48.10	2 4 0	21		-123	6.2	55.5						
Methyl methacrylate:	100.12	2 3 2	24		-48	100	107.0	10.1	8.5				360

continued

Table Da-1: Selected Properties for Selecting Separation Processes (at 25°C unless noted)

	mol mass	NFPA H F S	Do	Cuc s s s	freezing temp. °C	boiling temp C	Molar volume	δ_p $(J/cm^3)^{0.5}$	δ_h	pKa	pKsp	Biodegr mg COD h.g	BOD/ COD	ΔH_m kJ/kg	ΔH_v kJ/kg
N-Methylmorpholine:	101.15				-66	115									296
1-Methylnaphthalene:	142.2	2 2 0			-22	244.6	139.4	9.5	0.0					34	324
2-Methylnaphthalene:	142.2				34	241								83	324
2-Methylnonane:	142.28				-74.3	166.8	195.4	0.0	0.0						
3-Methylnonane:	142.28				-84.6	167.8	193.5	0.0	0.0						
4-Methylnonane:	142.28				-101.6	165.7	194.3	0.0	0.0						
5-Methylnonane:	142.28				-86.5	165.1	194.2	0.0	0.0						
2-Methyloctane:	128.26				-80.1	142.8	180.5	0.0	0.0						
3-Methyloctane:	128.26				-107.6	143	179.1	0.0	0.0						
4-Methyloctane:	128.26				-113.2	142.4	178.2	0.0	0.0						
2-Methyl oleate:	296.5				-20		337.3	3.9	3.7						
2-Methylpentanal	100.2						124.5	8.0	7.1						
2-Methylpentane:	86.18	1 3 0			-153.7	60.3	131.9	0.0	0.0					73	323
3-Methylpentane:	86.18	1 3 0				63.3	129.7	0.0	0.0						326
2-Methyl-1,3-pentanediol:	118.2	2 1 0					122.0	10.9	18.6						
2-Methyl-1,5-pentanediol:	118.2						121.9	11.6	20.6						
2-Methylpentanoic acid:	116.16	0 1 0				199-201	126.5	8.9	11.8	4.8					
2-Methylpentanol:	102.18	0 2 0				148	123.7	8.5	12.6						
2-Methyl-2-pentanol:	102.18				-103	120-22	122.4	9.2	11.1						
2-Methyl-3-pentanol:	102.18					126.7	124.0	8.0	12.1						
3-Methylpentanol:	102.18					152.4	124.0	8.6	13.0						
3-Methyl-2-pentanol:	102.18					134.3	123.0	8.2	12.6						
3-Methyl-3-pentanol:	102.18				-23.6	122.4	123.3	8.7	10.1						
4-Methylpentanol:	102.18					151.6	125.7	8.7	13.4						
4-Methyl-2-pentanol:	102.18														
2-Methyl-2-pentenal:	98.14					136-37	114.4	9.6	7.7						
2-Methyl-1-pentene:	84.16	1 3 0			-135.7	60.7	123.8	3.9	3.2						
2-Methyl-2-pentene:	84.16	1 3 0			-135	67.3	122.6	3.8	3.4						
3-Methyl-1-pentene:	84.16				-153	51.1	126.1	3.7	2.9						
4-Methyl-1-pentene:	84.16	1 3 0			-153.6	53.9	126.7	3.7	3.0						
cis-4-Methyl-2-pentene:	84.16	1 3 0			-134.4	56.3	125.8	3.5	0.0						
trans-4-Methyl-2-pentene:	84.16				-140.8	58.5	125.9	3.3	0.0						
N-Methylpiperazine:	100.16	2 2 0				138	111.5	9.0	5.2	4.94(+2)					
2-Methyl-2-propanethiol:	90.18				1	64								27.5	315
Methyl propionate:	88.11	1 3 0			-88	79	96.3	8.8	8.0						376
Methyl propyl ether:	74.12				-150	39									
Methyl propyl ketone:	86.13	2 3 0			-77.8	102	106.5	8.8	7.0						
N-Methylpyrrolidone:	99.13				-24	202	97.2	10.4	13.5						
N-Methyl-2-pyrrolidone:	99.13	2 1 0			-23	202	96.6	12.3	7.2						
Methyl salicylate:	152.15	1 1 0			-8	222	129.6	8.0	12.3						307
a-Methylstyrene:	118.18	2 2 0	14		-24	165-69	130.0	8.5	0.0						
2-Methyltetrahydrofuran:	86.13	2 3 0				78									
4-Methyltetrahydropyran:	100.16					109	105.1	6.3	0.0						
2-Methylthiophene:	98.17				-63	113	96.8	12.1	2.6					97	346
3-Methylthiophene:	98.17				-69	114	96.6	12.1	1.8					107	349
Methyl valerate:	116.2					128									
Methyl vinyl ketone:	70.09	2 3 2	24			81	83.7								
Monoethanolamine: MEA	61.08	2 1 0	4		10.3	170	60	13.8	25					336	816
Morpholine:	87.12	2 3 0		0 0 0	-4.7	128.3	87.1	11.4	10.1	8.5(+1)					425
N*															
Naphtha: petroleum:		1 3 0	16	0 0 0											300
Naphthalene:	128.17	2 2 0	14	1 0 0	80.5	218	128.6	2.0	5.9					149	337
1-Naphthol:	144.17				95	278									

2-Naphthol:	144.17				122	285						
Neon:	20.18				-248.5	-245.91						
Nicotinic acid:	123.11				237							
Nitric acid:	63.02	2 0 1		3 3 3	-42	86	41.96			-1.38		
90%wt:					-60	102						
60%wt:					-22	120						
Nitric oxide:	30.01				-163.8	-151.8						460
p-Nitroaniline:	138.13	3 1 3			146	390						
Nitrobenzene:	123.11	3 1 2	24	0 0 0	5-6	210-11	102.9	14.0	0.0		94	410
2-Nitrobiphenyl:	199.2	2 1 0			36-8	320						
4-Nitrobiphenyl:	199.21	2 1 1	14		112-4	340						
o-Nitrochlorobenzene:	157.56	3 1 1	29	1 2 1	44							
p-Nitrochlorobenzene:	157.56	2 1 3	29		83							
Nitroethane:	75.07	1 3 2	24	0 0 0	-89.5	115	71.8	13.0	0.0		131	506
Nitrogen:	28.01			0	-209.9	-195.8						199
Nitrogen dioxide:	46.01				-11.2	21.2	32					414
Nitrogen peroxide:	92.02				-11	21						
Nitroglycerine:	227.09	2 1 4	40	0 2 0	13.3	270 expl						
Nitromethane:	61.04	1 3 4	40	0 0 0	-29	100.8	53.7	18.8	5.1	10.2	159	577
2-Nitrophenol:	139.1				44	214						
4-Nitrophenol:	139.1				112	279						
1-Nitropropane	89.09	1 2 3	29	0 0 0	-108	130-31	89.3	11.2	0.0	8.98		432
2-Nitropropane:	89.09	1 2 3			-93	120	90.2	10.4	6.6			414
4-Nitrosodiphenylamine:	198.2				144 d							
N-Nitrosodipropyl amine:	130					206						
o-Nitrotoluene:	137.13	2 1 4		0 0 0	-4	222.3	117.9					
m-Nitrotoluene:	137.13	2 1 4			15-6	230-1	118					
p-Nitrotoluene:	137.13	1 1 3	29		51.9	237.7	120					
Nitrous oxide:	44.01			1 0 1	-90.8	-89.5	59.5					376
Nonadecane:	268.5				31.9	330.6	345				125	210
Nonadecanoic acid:	298.5				68.9	299(14kP	340				238	
n-Nonane:	128.26	0 3 0			-51	150.8	178.7	0.0	0		121	288
Nonanoic acid:	158.24				12.5	255.6	174.5			4.95	128	
2-Nonanone:	142.24	0 2 0			-7.5	195.3	157.4	6.9	4.6			
3-Nonanone:	142.24					187						
4-Nonanone:	142.24				...	187						
5-Nonanone:	142.24				-50	188						
1-Nonene:	126.24	0 3 0			-81	146	172.9	3.4	0		143	288
n-Nonyl alcohol:	144.26	1 2 0			-5.5	213.5	174.4	7.3	12.0			377
2-Nonyl alcohol: (2-nonanol)	144.26				-36--35	193-94	174.4	6.6	10.3			
n-Nonylbenzene:	204.36	0 1 0				280.1	238.1	6.8	0.0			
1-(n-Nonyl)naphthalene:	254.4	0 2 0					273.0	6.6	0.0			
Nonyl phenol:	220.36	2 1 0					235.2	6.7	9.1			

O*

Octacosanoic acid:	424.8				92-4							
n-Octadecane:	254.5				28.2	316.7	327.9				244	215
Octadecanoic acid:	284.48	1 1 0	4	3 3 0	71.2	376.1	302.4	3.3	5.5		199	234
1-Octadecanol:	270.5				57.9							
1-Octadecene:	252.49				17.7	314.9					130	212
9-Octadecenoic acid:	282.47	0 1 0	4	3 3 1	16.3	222	316.1	3.1	5.5			240
9-Octadecen-1-ol:	268.48				6-7	195(1kPa	316.3	2.7	8.0			
Octadecylamine:	269.5				50	232(4kPa						
Octane:	114.23	0 3 0	16		-56.8	125.7	162.6	0.0	0.0		182	301
n-Octanoic acid:	144.21				16.5	239.3	158.7	3.3	8.2	4.9	96	486
1-Octanol:	130.23	1 2 0		0 0 0	-16.7	194.4	157.5	7.7	12.4		324	408
2-Octanol:	130.23	1 2 0		0 0 0	-31.6	180	159.0	6.9	10.9			395

continued

Table Da-1: Selected Properties for Selecting Separation Processes (at 25°C unless noted)

	mol mass	NFPA H F S	Do	Cuc s s s	freezing temp. °C	boiling temp C	Molar volume	δ_p $(J/cm^3)^{0.5}$	δ_h	pKa	pKsp	Biodegr mg COD h.g	BOD/ COD	ΔH_m kJ/kg	ΔH_v kJ/kg
1-Octene:	112.22	1 3 0			-101.7	121.3	157.0	3.5	2.3					137	306
cis-2-Octene:	112.22				-100.2	125.6	154.9	3.2	0.0						
trans-2-Octene:	112.22				-87.7	125	155.9	3.0	0.0						
cis-3-Octene:	112.22				-126	122.9	156.1	3.2	0.0						
trans-3-Octene:	112.22				-110	123.3	156.9	3.0	0.0						
cis-4-Octene:	112.22				-118.7	122.5	155.6	3.2	0.0						
trans-4-Octene:	112.22				-93.8	122.3	157.1	3.0	0.0						
n-Octylbenzene:	190.3						223.6	5.4	0						
tert-Octyl mercaptan:	146.3	2 2 0	10			197									
							IV	SV	AV						
Oils, crude:		1 3 0	16												
Oils, edible, canola:					-10		98	173	0.7						
Oils, edible: castor:					-12		84	180	0.5						
Oils, edible, coconut:					14-22		6-10	200	5						
Oils, edible, corn:					-20--10		111-28	190	1.4						
Oils, edible, cottonseed:					0		108	195	0.7						
Oils, edible, fish:					-5		160	192	7						
Oils, edible, olive:		0 1 0			-6		80	193	0.8						
Oils, edible, palm:		0 1 0			21		53	200	10						
Oils, edible, peanut:		0 1 0			3		93	194	0.8						
Oils, edible, safflower:					-15		130	197	0.6						
Oils, edible, soya bean:					-20		128	193	0.9						
Oils, edible, vegetable:															
Oils, linseed:					-19		200	190	2						
Oils, lubricating:															
Oils, mineral:		0 1 0	4												
Oils, sperm:					15.5		83	130	13						
Oxalic acid:	90.04			1 3 3	190	s				1.23			0.89		
Oxalic acid-dihydrate:	126				104-6		76								
1-4-Oxathiane:	104.17					147	93.5	12.3	5.2						
Oxygen:	32				-218.4	-183									213
P*															
Paraformaldehyde:					156d										
Paraldehyde:	132.16	2 3 1		0 0 0	12.6	124.5								105	
Pentachloroethane:	202.33				-29	160								56	184
Pentachlorophenol:	266.34				190	310d						A			
n-Pentadecane:	212.42				9.9	270	276.2	0.0	0					164	234
Pentadecanoic acid:	242.4				52.3	339.1	287.8								
Pentadecanol:	228.42				35	299									240
1,2-Pentadiene:	68.12				-137	45									424
trans-1-3-Pentadiene:	68.12				-87.5	42	100.7	5.3	4.8						398
1,4-Pentadiene:	68.12				-149	26								91	370
2,3-Pentadiene:	68.12				-126	46									415
Pentamethylbenzene:	148				54	231								83	
Pentane:	72.15	1 4 0	21	1 1 2	-130	36.1	115.2	0.0	0.0					117	358
1,5-Pentanediol:	104.15	1 1 0			-18	260	104.8	12.2	22.8						
Pentanoic acid:	102	2 1 0			-34.5	187	108.4	9.7	14.3	4.8				139	432
Pentatriacontanoic acid:	522.94														
1-Pentene: (Amylene)	70.13	1 4 0			-138	30	109.5	4.1	0.0					83	360
cis-2-Pentene:	70.13				-151.4	36.9	107.0	3.8	0.0					101	373
trans-2-Pentene:	70.13				-136	36.3	108.2	3.4	3.6					111	373
3-Pentyl alcohol:	88.15					116.1	107.3	8.3	13.0						495
Pentylbenzene:	148.24				-78.2	231									

n-Pentylcyclohexane:	154.3				-57.5	202.8	192.0	2.3	0.0				
Pentylcyclopentane:	140.28				-92	180							
Pentyl mercaptan:	104.2				-75.7	126.6						168	337
1-Pentyne:	68.11				-105.7	40.2	98.7						408
Peracetic acid:	76.05	3 2 4	40	? ? 0	0.1	105expl							
Perfluorodimethylcyclohexane:	400.06				-55	101-02	218.9	0.0	0.0				
Perfluoromethylcyclohexane:					-37	76	195.8	0.0	0.0				
Perfluoro-n-heptane:							227.3	0.0	0.0				
Phenanthrene:	178.23				100	340						102	312
o-Phenetidine:(2-)	137.2				<-20	232.5							
m-Phenetidine:(3-)	137.2					248							
p-Phenetidine:(4-)	137.2				2.4	254							
Phenetole:	122.16				-29	170							337
Phenol: (Carbolic acid)	94.11	3 2 0	4	2 2 3	43	181.7	89.0	5.9	14.9	10	80	121	487
Phenyl acetate:	136.15					196							
N-Phenylethanolamine:	137.18	2 1 0			56-57		125.7	11.9	17.0				
Phenyl ether:	170.2				27	259						101	277
Phenylhydrazine:	108.14	3 2 0		0 3 ?	19.8	243	98.4	14.9	13.7	5.2(+1)			
N-Phenylpiperazine:	162.23					286.5	152.7	11.4	3.2	8.7(+1)			
Phosgene:	98.92	(4)		3 3 3	-118	7.6	71.6					58	239
Phosphoric acid-	98				42								
-115%wt H3PO4:					38.85								
-105%wt H3PO4:													
-85%wt H3PO4:	98			3 3 3	21	530							
-75%wt H3PO4:					-20	300							
Phthalic acid:	166.13			1 3 0	208	d					78.4		
Phthalic anhydride:	148.11	2 1 0		2 1 0	141	284							
Phthalide:	134.13				72	290							
a-Picoline:	93.13	2 2 0	14		-66.8	128.8	98.6	7.8	6.9				380
b-Picoline:	93.13				-19	143.5							
gam-Picoline:	93.13	2 2 0			2.4	143.1							
Picric acid:	229.11	2 4 4		3 3 1	120-2	expl.							
Pinacolyl alcohol:	102.18				5.6	120.14	125.8	8.0	10.6				
Pinene:	136.24	1 3 0			-55	156.2	158.8	4.3	0.0				
Piperidine:	85.15	2 3 3			-9	106	98.9	8.7	5.8	2.9(+1)			374
Pivaldehyde:	86.13				6	77-78	108.7	8.5	5.5				
Polyglycolamine H221M	221.3						220.4	9.2	10.8				
Potassium:	39.1	3 1 2	24		63.4	759							
Potassium benzoate:	160.22				>300								
Potassium chloride:	74.56				790	1500							
Potassium hydroxide:	56.1	3 0 1			380	1320							
Potassium nitrate:	101	1 0 2	29		333	400d							
Potassium sulfate:	174.25				tr. 588								
Potassium xanthate:	160	2 1 0				200 d							
Propadiene: (Allene).	40.07				-136	-34							506
Propane:	44.1	1 4 0	21	0 0 0	-189.7	-42.1	89.6	0.0	0.0			80	426
1,3-Propanediol:	76.1	1 - 0			-27	214	72.1						759
2-Propanethiol:	76.16				-131	57							
Propanoic acid:	74.08	2 2 0		0 3 3	-20.8	141	74.6	11.2	17.1	4.87		102	440
1-Propanol:	60.10	1 3 0	16	0 0 0	-126.5	97.4	74.8	10.5	17.7		71	89	688
b-Propiolactone:	72				-33.4	162d							
Propionaldehyde:	58.08	2 3 1			-81	48.8	72.1	9.7	11.0				486
Propionic anhydride	130.14	2 2 1			-43	167	128.2	11.4	10.4				348
Propionitrile:	55.08	4 3 1	16		-92.9	97.3	70.5	11.5	13.8			92	607
Propyl acetate:	102.13	1 3 0	16	0 0 0	-95	102	115	8.1	7.8				336
n-Propylamine:	59.11	3 3 0	16		-83	47.8	82.4	9.3	7.2	10.5(+1)			
n-Propylbenzene:	120.19	2 3 0	16		-99.5	159.2	139.4	6.9	0.0			17	318
n-Propyl chloride:	78.54	2 3 0	16		-122.8	46.6	88.2	7.2	6.8			71	347
n-Propylcyclohexane:	126.24				-94.9	156.7	159.1	2.5	0.0			82	286

continued

Table Da-1: Selected Properties for Selecting Separation Processes (at 25°C unless noted)

	mol mass	NFPA H F S	Do	Cuc s s s	freezing temp. °C	boiling temp C	Molar volume	δ_p (J/cm^3)$^{0.5}$	δ_h	pKa	pKsp	Biodegr mg COD h.g	BOD/ COD	ΔH_m kJ/kg	ΔH_v kJ/kg
n-Propylcyclopentane:	112.2						145.5	3.1	0.0					89	304
Propylene: (Propene)	42.08	1 4 1	21	0 0 0	-185	-48	69.							71	438
Propylene-1-2-carbonate:	102.09	1 1 0			-55	240	85.9	18.0	4.1						
sec-Propylene chlorohydrin:	94.54	2 2 0				126-27	84.8	11.6	16.0						
Propylenediamine:	74.1	2 3 0					86.4	12.8	12.8						
Propylene dichloride:	113.0	2 3 0	16	0 1 1	-99.5	120.4	95.1	10.4	5.4						
1,2-Propylene glycol:	76.1	1 1 0	4	0 0 0	-60	189	73.4	14.1	27.1						688
Propylene glycol-															
-acrylate:(mono)	130														
-butyl ether:(mono)	176.3				-100	230									142
-dimethyl ether:	104.2				-28	96									
-ethyl ether:(mono)	104.2				-100	132.8									281
-methyl ether:(mono)	90.12				-80	121	98.3	9.2	14.3						386
-methyl ether acetate:	132.16					145									
-phenyl ether:(mono)	152.2				13	240									
-isopropyl ether:(mono)	118					48									
-propyl ether:(mono)	118				-80	150									
Propyleneimine:	57.09					66									
Propylene oxide:	58.08	2 4 2	21	1 1 0	-112	34	70.0	8.2	10.4					113	481
Propyl formate:	88	2 3 -			-92.9	80.85									369
4-Propylheptane:	142.3						194.7	0.0	0.0						
2-Propylheptyl alcohol:	158.3						191.2	6.9	10.6						
Propyl mercaptan:	76.16				-113	67.6									416
Propyl propionate:	116.16				-75.7	123									
m-Propyltoluene:	134.2						157.0	6.3	0.0						
o-Propyltoluene:	134.2						154.6	6.5	0.0						
p-Propyltoluene:	134.2						157.3	6.2	0.0						
Propyl valerate:	144.2				-71	68									
Pseudocumene:	120.19	0 2 0			-43.8	169.3	137.2	7.1	0.0					103	327
Pyrene:	202.26				156	404						D			
Pyridine:	79.1	2 3 0	24	1 1 0	-42	115.5	77.5	10.1	7.7	8(+1)			1.2	94	460
2-Pyrrolidone:	85.11	2 1 0			24.6	245	76.0	17.4	11.3						
Q*															
Quinoline:	129.16	2 1 0		0 2 0	-16--15	238	118.2	7.0	7.6	4.8(+1)				84	386
R*															
S*															
Salicylaldehyde:	122.12	0 2 0			-7	196.7				8.14					314
Salicylic acid:	122				1	196									
Silane:	32.09				-185										
Silicon:	28.06				1420	2600								1105	15957
Silicon dioxide:	60.06					2230									
Silicon tetrachloride:															
Sodium:	22.99	3 1 0	24		97.5	880									
Sodium acetate:	82				324										
Sodium benzoate:	144.11														
Sodium bicarbonate:	84														
Sodium bromide:	102.9				755	1390									
Sodium chlorate:	106.45	1 0 2	24	0 3 1	248	d									
Sodium chloride:	58.45				801	1465	27							480	
-brine: 25%															
Sodium dodecyl sulfate:	288.3														

Sodium fluoride:	42				992							
Sodium hydrogen phosphate:												
Sodium hydrosulfide:	92.1	3 1 2	24	? 2 0	d							
Sodium hydroxide:	40	3 0 1		3 3 3	318	1390						
sodium hydroxide:50%												
Sodium nitrate:	69	1 0 2			271	320d						
Sodium nonyl sulfate:	246.2											
Sodium sulfate:	142.05											
Styrene:	104.15	2 3 2	24	1 0 0	-30.6	145.2	115	1.0	4.1		105	372
Styrene oxide:	120.15	2 2 0			-35.6	194.1	114	11.4	5.1			
Succinic anhydride:	100.7				26.1	119-20	66.8	19.2	16.6		204	
Succinonitrile:	80.1				57.88	267					46	607
Sucrose:	342.3				192 d							
Sulfanilamide:	172.21				165							
Sulfathiazole:	255.32				200							
Sulfobromophthalein:	838											
Sulfolane:	120.17	2 1 0			27.7	285					11.4	487
Sulfur:	64	2 1 0	4		113	444						
Sulfur dioxide:	64	2 0 0	1	3 3 1	-73	-10	46.7				116	387
Sulfur hexafluoride:	146				-50							
Sulfuric acid:	98			3 3 3	10.5	340d				-3		
110%												
98%												
60%												
Sulfur trioxide:	80			1 3 1	16.8	44.8	42.6				108	540
T*												
Tallow:					31-38							
a-Terpineol:	154.25				40-41	220	165.2	7.9	10.2			
1,1,2,2-Tetrabromoethane:	345.67				0	243.5	116.6	5.1	8.2			
Tetrachlorodifluoroethane:	203				26	92.8						172
Tetrachlorodiphenyl:	292				-7	340						
1,1,2,2-Tetrachloroethane:	167.85				-36	146.2	105.2	5.1	9.4			230
Tetrachloroethylene:	165.83			0 0 0	-19	121	102.2	6.5	2.9		64	210
Tetracosanoic acid:	368.65				84.2		449.5					
Tetradecane:	198.4				5.9	253.5	260					
Tetradecanoic acid:	228.38				53.9	326.2					196	
1-Tetradecene:	196				-12.9	251						239
Tetradecylbenzene:	274				16	359						221
1,1,3,3-Tetraethoxypropane:	220.31	0 2 0				220	239.7	7.4	6.7			
Tetraethylene glycol:	194.23				4	327						
-diacrylate:	302											
-dibutyl ether:	306	2 1 0				335						
-diethyl ether:	250											
-dimethyl ether:	222.3	1 1 0			-30	275	220.8	8.4	10.3			
-methyl ether:(mono)	208.26					166(1.5kPa)						
Tetraethylene pentamine:	189				-30	340						
1,2,3,6-Tetrahydro-												
-benzaldehyde:	110.16	2 2 0				163-64	117.2	10.3	7.6			
Tetrahydofuran:	72.11	2 3 0			-108	67	81.1	11.0	6.7			410
1,2,3,4-Tetrahydro-												
-naphthalene:	132.21	1 2 0			-35.8	207.6	136.3	7.8	0.0			320
Tetrahydropyran:	86				-45	88						
Tetrahydropyran-2-methanol:	116.16	1 2 0				187	113.1	9.6	13.5			
Tetrahydrothiophene:	88.17				-96.2	120.9	88.2					
1,1,3,5-Tetramethoxyhexane:	206.3						216.7	7.7	7.1			
Tetramethyl-												
-ammonium chloride:	109.6				>230							

continued

Table Da-1: Selected Properties for Selecting Separation Processes (at 25°C unless noted)

	mol mass	NFPA H F S	Do	Cuc s s s	freezing temp. °C	boiling temp C	Molar volume	δ_p (J/cm^3)$^{0.5}$	δ_h	pKa	pKsp	Biodegr mg COD h.g	BOD/ COD	ΔH_m kJ/kg	ΔH_v kJ/kg
-benzene, (1,2,3,4):	134.22				-6.2	205								84	336
-benzene, (1,2,3,5):	134.22				-24	198								80	327
-benzene, (1,2,4,5):	134.22				79	197								157	339
2,2,3,3-Tetramethylbutane:	114.23				-120	106-07	139.3	0.0	0.0					62	277
N,N,N,N-Tetramethyl-															
-1,3-butadiamine:	144.3					169	181.1	4.7	2						
-1,2-ethylenediamine:	116.21				-55	120-22	150.9	5.1	2.7						
2-2-3-3-Tetramethylpentane:	128.26	0 3 0				140.27	171.9	0.0	0.0						
N,N,N,N-Tetramethyl-															
-1,3-propanediamine:	130.24					145-46	167.2	4.8	2.6						
Tetramethyl urea:	116.16				-1.2	166-67	119.9	8.2	11.1						
2-Thiabutane:	76.2				-105.9	66.7	91.3	8.7	5.7					128	388
2-Thiapropane:	62.1						73.8	9.6	6.7					129	435
Thiodiglycol:	122.19				-16	282									
Thiophene:	84.14	2 3 0			-38.2	84.2	79.0	12.4	7.5					59	375
Thiophosgene:	114.98					73									300
Thiourea:	76.12					175									
Thymine:	126.12				316	subl.									
Toluene:	92.14	2 3 0	16	3 0 0	-95	110.6	106.3	8.0	1.6				0.6	72	364
Toluene-2,4-diisocyanate:	174.16	2 1 1				134	142.2	16.4	7.8						
m-Toluidine:	107.16				-30	203								36	419
o-Toluidine:	107.16				-15	199									417
p-Toluidine:	107.16				43	200								177	414
Triallylamine:	137.22			3 0 ?		155-56	169.6	6.6	4.1	8.3(+1)					
Triacontanoic acid:	452.8					91-92									
1,2,4-Triazole:	69				119	260									
Tributylamine:	185.35	2 2 0	4	3 0 0	-70	213	238.2	2.8	4.0						
Tri-n-butyl phosphate:	266.32	2 1 0		? 0 ?	-79	289 d	272.0	16.4	6.3						
1,2,4-Trichlorobenzene:	181.45	2 1 3	29		16	214	124.8								
1,1,1-Trichloroethane:	133.4	2 1 2	24	0 0 0	-30.4	74.1	99.6	4.3	2.1					34	250
1,1,2-Trichloroethane:	133.4			0 0 0	-36.5	113.8	92.7	12.9	7.0					85	260
Trichloroethylene:	131.39	2 1 1	14	3 3 1	-73	87	89.7	3.1	5.3						240
Trichlorofluoromethane:	137.37				-110	23.7	91.9	2.0	0.0						
2,4,6-Trichlorophenol:	197.45				64	246									
1,2,3-Trichloropropane:	147.43	3 2 0		0 0 0	-14.7	156.8	106.1	12.7	6.7						252
2,2,3-Trichloro-															
-propionaldehyde:	161.42						110.0	14.6	8.0						
1,2,4-Trichlorotoluene:	195.48				-5	221									
1-1-2-Trichloro-															
-trifluoroethane:	187.38			0 0 0	-36.4	47.7	119.0	1.6	0.0						
Tricresyl phosphate:	368.37	2 1 0		3 3 1		265	322.3	12.3	4.5						186
n-Tridecane:	184.37				-5.5	235.4	243.7	0.0	0.0						
1-Tridecanol:	200.36	0 1 0			32-33		243.7	3.1	9.0						270
1-Tridecene:	182.35				-23	232									250
Triethanolamine:	149.19	1 1 1	14	3 0 0	20-1	277-9	132.5							182	410
Triethylamine:	101.19	2 3 0	16	3 0 0	-114.7	89.3	139.1	3.7	1.9	10.7(+1)					318
1,3,5-Triethylbenzene:	162.28					215									270
Triethylene glycol:	150.17	1 1 0	4	0 3 0	-5	278.3	133.5	12.2	21.8			27.5			410
-butyl ether:(mono)	206.28				-48 d										409
-diacetate:	234.24				-61	11.7									244
-diethyl ether:	206						217.2	7.5	8.8						
-dimethyl ether:	178.2	1 1 0			-45	3.8									336
-ethyl ether:(mono)	178.2	0 1 0			-18.7	256		8.8	13.5						264
-methyl ether:(mono)	164.2	0 1 0			-44	249	157.4	9.4	14.8						328
Triethylenetetramine:	146.24	3 1 0		3 0 0	12	266-27	148.9	12.4	14.1						

Triethyl phosphate:	182.16	0 1 1		0 0 ?			169.9	11.5	9.2		
1,1,1-Trifluoroethane:	84.04				-111.3	-47.3					
Trifluoromethane:	70.01				-155.2	-82.2					
Triglycol dichloride:	187.07	2 1 0				235	156.3	10.5	9.6		
1,3,5-Triisopropyl benzene:	204.36	2 3 1	16			232-6	241.8				
1,1,3-Trimethoxybutane:	148.2						161.8	7.8	6.6		
Trimethylamine:	59.11	2 4 0	21	3 0 0	-117.2	2.9	93.0				
1,2,3-Trimethylbenzene:	120.2	0 2 0		? 0 0	-25	175-76	134.5	7.3	0.0		
2,2,3-Trimethylbutane:	100.21	0 3 0			-25	80.9	145.2	0.0	0.0		
1,1,2-Trimethylcyclopentane:	112.22					113-14	146.3	3.0	0.0		
1,1,3-Trimethylcyclopentane:	112.22					115-16	145.7	2.9	0.0		
2,2,3-Trimethylheptane:	142.3						193.1	0.0	0.0		
2,2,4-Trimethylheptane:	142.28					147.7	195.6	0.0	0.0		
2,2,5-Trimethylheptane:	142.3						197.4	0.0	0.0		
2,2,6-Trimethylheptane:	142.3						199.3	0.0	0.0		
2,3,3-Trimethylheptane:	142.3						191.5	0.0	0.0		
2,3,4-Trimethylheptane:	142.3						190.8	0.0	0.0		
2,3,5-Trimethylheptane:	142.3						193.3	0.0	0.0		
2,3,6-Trimethylheptane:	142.3						195.2	0.0	0.0		
2,4,4-Trimethylheptane:	142.3						195.5	0.0	0.0		
2,4,5-Trimethylheptane:	142.3						193.3	0.0	0.0		
2,4,6-Trimethylheptane:	142.3						198.5	0.0	0.0		
2,5,5-Trimethylheptane:	142.3	0 2 0				151	194.7	0.0	0.0		
3,3,4-Trimethylheptane:	142.3						189.0	0.0	0.0		
3,3,5-Trimethylheptane:	142.3					155.7	196.3	0.0	0.0		
3,4,4-Trimethylheptane:	142.3						189.2	0.0	0.0		
3,4,5-Trimethylheptane:	142.3						188.7	0.0	0.0		
2,2,3-Trimethylhexane:	128.26					131.7	177.2	0.0	0.0		
2,2,4-Trimethylhexane:	128.26				120	126.5	180.4	0.0	0.0		
2,2,5-Trimethylhexane:	128.26	2 3 0			-105.8	124	181.4	0.0	0.0	48	264
2,3,3-Trimethylhexane:	128.26				-116.8	137.7	175.0	0.0	0.0		
2,3,4-Trimethylhexane:	128.26				-127.9	139	174.7	0.0	0.0		
2,3,5-Trimethylhexane:	128.26				-123.4	131.3	164.1	0.0 0.0			
2,4,4-Trimethylhexane:	128.26					126.5	180.4	0.0	0.0		
3,3,4-Trimethylhexane:	128.26				-101.2	140.5	173.3	0.0	0.0		
2,6,8-Trimethyl- -4-nonanol:	186.3	2 2 0				225	228.8	5.9	7.8		
2,2,3-Trimethylpentane:	114.23	0 3 0		0 0 0	-112.3	110	159.5	0.0	0.0	76	281
2,2,4-Trimethylpentane:	114.23	- 3 0		0 0 0	-107.4	99.2	165.1	0.0	0	81	272
2,3,3-Trimethylpentane:	114.23	0 3 0		0 0 0	-100.7	114.7	157.3	0.0	0	13	285
2,3,4-Trimethylpentane:	114.23			0 0 0	-109.2	113.4	158.9	0.0	0	81	286
2,2,4-Trimethylpentanol:	130.23					168-9	159.2	7.7	9.9		
Trimethyl phosphate:	140.08			? 3 0		197	117	16	10.2		
2,4,6-Trinitrotoluene:	227.13				80.1	explo					
Triphenylmethane:	244.34	0 1 0			93.4	360					
Tripropylene glycol:	192.3				-45	267.2					
-butyl ether:	248.4					276					
-ethyl ether:	220.3					486					
-isopropyl ether:	234.8					112.7					
-methyl ether:	206	0 1 0			-42	242.4					
Tryptophan:	204.23				280 d						
U*											
Uracil:	112.09				338						
Undecane:	156.31				-25.6	196	211.2	0.0	0	143	266
Undecanoic acid:	186.3				29.3	284				135	
1-Undecene:	154.3				-49	192				110	265
n-Undecyl alcohol:	172.31				19	243	207.7	6.7	11.2		

continued

Table Da-1: Selected Properties for Selecting Separation Processes (at 25°C unless noted)

	mol mass	NFPA H F S	Do	Cuc s s s	freezing temp. °C	boiling temp C	Molar volume	δ_p $(J/cm^3)^{0.5}$	δ_h	pKa	pKsp	Biodegr mg COD h.g	BOD/ COD	ΔH_m kJ/kg	ΔH_v kJ/kg
2-Undecyl alcohol:	172.31				0	228	209.4	6	9.8						
Uranium oxide:	842.21				d										
Urea:	60.06				132.7	d	458								1464 s
Urea nitrate:	123				152	d									
Urethane:	89.09				48	182									
V*															
Valeraldehyde:	86.13	1 3 0			-91.5	103	106.4	8.4	8.8						
Valine:	117.15				>300										
Veratrole:	138.17				15	206									
Vinyl acetate:	86.09	2 3 2	16		-93.2	72.2	92.4	10.0	8.7						360
Vinyl allyl ether:	84.1	2 3 2	24				105.4	7.9	7.0						
Vinyl butyl ether:	100.2	2 3 2	24				129.5	6.1	5.5						322
Vinyl butyrate:	114.2	2 3 2					127.6	8.7	7						
Vinyl chloride:	62.5	2 4 1	21		-154	-13	68.8							76	333
Vinyl 2-chloroethyl ether:	106.6	2 3 2					102.4	9.9	8.7						
Vinyl crotonate:	112.13	2 3 2				133-4	119.3	10	8.4						
Vinyl ethyl ether:	72.1	2 4 2	21				96.4	7.0	5.7						372
Vinyl 2-ethylhexanoate:	170.25	2 2 2			-90		194.6	7.0	5.2						
Vinyl 2-ethylhexyl ether:	156.3	2 2 2					194.2	5.2	4.5						
Vinyl isobutyl ether:	100.2	2 3 2					131.3	6.3	4.0						
Vinyl isopropyl ether:	86.1	2 4 2					115.3	6.6	4.2						
Vinyl methyl ether:	58.1	2 4 2					78.3	7.0	6.8						
Vinyl propionate:	100.12	2 3 2				91-92	110	9.4	7.7						
m-Vinyltoluene:	118.18	2 2 1	14		-82	170-1	130.9	8.3	0.0						
o-Vinyltoluene:	118.18					171	129.8	8.4	0						
p-Vinyltoluene:	118.18				-37.8	170-5	134.9	8.2	0						
W*															
Water:	18.016				0	100	18.1	22.8	40.4	15.74				334	2263
Water, heavy:	20.23				3.82	101.42									
X*															
Xenon:	131				-112	-108									
m-Xylene:	106.17	2 3 0	16		-47.9	139.1	122.9	7.2	2.4					109	343
o-Xylene:	106.17	2 3 0	16		-25.2	144.4	118.4	7.0	0.0					128	347
p-Xylene:	106.17	2 3 0	16		13.3	138.3	123.3	7.0	2.0					161	340

Table Db-1 Selected Physical and Termal Properties of Gases and Liquids for Sizing Equipment (at 25°C unless noted)

	Formula	LIQUID								GAS						References
		dens Mg/m	dD/dT x1000	vis mPa.s	dv/dT x1000	cp kJ/kg.K	dc/dT x1000	k mW/m.K	dk/dT x1000	vis μPa.s	dv/dT x1000	cp kJ/kg.K	dc/dT x1000	k mW/mK	dk/dt x1000	
Abietic acid:	C20H30O2	3.97	-2.9	27.6	n=-9.7	1.76	3.8	143		4.2	16			8.2	50	4
Acenaphthene:	C12H10	1.069		4.1	n=-4.1	1.49	3.1	136	-84	5.1	18			8.4	54	4
Acetal:	C6H14O2	0.8213	-0.103			0.5199										14,4
Acetaldehyde:	C2H4O	0.771	-1.7	0.21	n=-3.4	2.19	3.33	171	-378	8	27	1.26	2.6	12.6	42	1,13,4
Acetamide:(ethanamide):	C2H5NO	1.159		4	-36											13,14,4
Acetanilide:	C8H9NO	1.219		5.26	n=-6.2	1.53	2.9	147		6	21			8.4	52	14, 15,4
Acetic acid:	C2H402	1.044	-1.1	0.81	n=-4.4	2.05	3.4	171	-180	8	37	1.44	1.5	10.5	71	1,13,14,4
		0.939*		0.372*		2.42*		158*		10.4*		1.39*		20.7*		
Acetic anhydride:	C4H6O3	1.075	-1	0.76	-7	1.82	6.2	221	-70	7	26	1.26	2	10	84	13,14,4
Acetone:	C3H6O	0.785	-1.2	0.33	-2	2.17	2.8	169	-260	7	24	1.51	2.8	11.2	70	13,14,4
Acetone cyanohydrin:	C4H7NO															4
Acetonitrile:	C2H3N	0.7803	-1.2	0.37	n=-2.85	2.42	2.5	180	-489	2.2	14	1.34	2	12	80	1,13,14,2,4
Acetophenone:	C8H8O	1.024	-0.45	1.7	n=-4.8	1.888	1.7	141	-67	12.8(300)	22	1.835(300	2.1	30(300)	100	13,14,4
1-Acetoxy-1,3-butadiene:	C6H8O2	0.947														15
Acetylacetone:	C5H8O2	0.968		0.773	n=-3.3	2.08	4	163	-296	6.1	22			10.4	79	4
Acetyl chloride:	C2H3OCl	1.105		0.75	-6.9	1.4651	1.67	150	-180	8.3	27	0.8372	1.3	9.2092	60	2,4
Acetylene:	C2H2	1.051	-5.5	3.1(-75)		3.1(-70)	2.4	540(-70)		10.3	32	1.76	1.9	20.4	90	1,13,4
		0.617*		1.69*		3.09*		55.9*		7.35*		1.47*		11.2*		
Acetyl peroxide:	C4H6O4															23,22
Acetylsalicylic acid:	C9H8O4	1.35														23,15
Acrolein:(2-propenal)	C3H4O	0.835	-1.1	0.431	-2.5	2.14	2.9	205	-753	8.2	27	1.15	2.5	10.9	60	2, 14,4
Acrylamide:	C3H5NO	1.122		7.26	n=-6.5	1.58	2.2	142	-221	7.2	26			9.4	68	4
Acrylic acid:	C3H4O2	1.04		1.3	n=-4.1	2.03	1.7	226	-125	8	28	1.047	2.7	8.372	180	2,4
Acrylonitrile:	C3H3N	0.801	-1.1	0.34	-3.6											4
Adipic acid:	C6H10O4	1.07(170)		4.54(160)		2.43						1.68(300)				4
Adiponitrile:	C6H8N2	0.9625		6.16	-7.2											4
Air:		0.07695	7	0.084	-5.9	2.04	8	111	-1760	17.9	48	1.007	0.07	25.7	80	13
Allyl acetate:	C5H8O2	0.922	-1.08	0.2068												4,15,14
Allyl acetoacetate:	C7H10O3	1.032														20
Allyl alcohol:	C3H6O	0.8464	-0.87	1.3	n=-5.4	2.785		180				2.43				14
Allyl amine:	C3H7N	0.7575	-1.07	0.3745												14
Allyl bromide:	C3H5Br	1.398		0.46	-3											15
Allyl chloride:	C3H5Cl	0.931	-0.69	0.32	n=-2.9							1.328				4
Allyl chloroformate:	C4H5O2Cl	1.13														15
Allyl cyanide:	C4H5N	0.83														15
Allyl iodide:	C3H5I	1.846		0.66	-4											15
2-Aminobiphenyl:	C12H11N															15,23
2-Amino-1-Butanol:	C4H11NO	0.947														15,23
Aminodiphenylamine:	C12H12N2															4
12-Aminododecanoic acid:	C12H25NO2															20
2-(2-Aminoethoxy)-ethanol:	C4H11O2N	1.46														15
N-(2-Aminoethyl)ethanolamine:	C4H12N2O	1.025														4,23,15
4-(2-Aminoethyl)morpholine:	C6H14N2O	0.992														23,20
N-(2-Aminoethyl)piperazine:	C6H15N3	0.978														23,22
Aminoheptane	C7H17N	0.766														
6-Amino-1-hexanol	C8H19NO															
2-Aminonaphthalene	C10H9N															
3-Amino-1-propanol	C3H9NO	0.982														
4-(3-Aminopropyl)morpholine:	C7H16N2O	0.981														23
Ammonia:	H3N	0.6	-1.7	0.13	n=-3.9	4.73	2	451	-1270	10.2	35	2.16	1.3	24.2	100	4
Ammonium acetate:	C2H7NO2	1.073														19
bicarbonate:	CH5NO3	1.573														19
chlorate:	ClH4NO3	1.8														19
chloride:	ClH4N	1.53														19
hydroxide:	H5NO															19
nitrate:	H4N2O3	1.66														19

continued

Table Db-1 Selected Physical and Termal Properties of Gases and Liquids for Sizing Equipment (at 25°C unless noted)

	Formula	LIQUID								GAS						
		dens Mg/m	dD/dT x1000	vis mPa.s	dv/dT x1000	cp kJ/kg.K	dc/dT x1000	k mW/m.K	dk/dT x1000	vis µPa.s	dv/dT x1000	cp kJ/kg.K	dc/dT x1000	k mW/mK	dk/dt x1000	References
oxalate:	C2H8N2O4.H2O	1.501														19
phosphate:	H6NO4P	1.08														19
sulfate:	H8N2O4S	1.769														19
p-Amyl acetate:	C7H14O2	0.8707	-0.997	0.82	n=-6	2.125		144								14
act-Amyl alcohol:	C5H12O	0.815		4	-45	2.5		164	-120							14,4
p-n-Amyl alcohol:	C5H12O	0.811	-0.86	3.347	-182	2.361	9.3	155	160	6.875	23	1.59	3.13	5.25	130	13,14,4
sec-Amyl alcohol:	C5H12O	0.8054	-0.799	3.58	-160											14,4
tert-Amyl alcohol:	C5H12O	0.805		3.7	-178	2.755										14,4
n-Amylamine:	C5H13N	0.8														4
Amyl bromide:	C5H11Br	1.22														15
Amyl butyrate:	C9H18O2	0.871														3
Amyl chloride:	C5H11Cl	0.878	-0.99	0.547	-4.3	1.785	2.9	119	-210			1.23				14
Amyl ether:	C10H22O	0.779		1.01	-17.7							1.58				14
Amyl formate:	C7H12O	0.882														23
4 tert-Amyl phenol	C11H16O															20
Anethole:	C10H12O	0.991				2.56										4
Aniline: (benzeneamine)	C6H7N	1.018	-93	3.77	n=-8.4	2.05	3	191	-190	68.5	280	1.17	4	10.4	113	13,14,4
		0.875*		0.303*		2.37*		154*		11.8*		1.74*		23.5*		
Anisidine:	C7H9NO	1.098														15
Anisole (methoxybenzene)	C7H8O	0.999	-0.932	1.1	n=-3.4	1.932										14,4
Anthracene:	C14H10	1.25														4
Anthraquinone:	C14H8O2															20
Arginine:	C6H14N4															20
Argon:	Ar	0.24	-4		-3.2	0.984		17.7	-1200	22.7	620	0.522	0	17.7	45	13
		1.393*		0.2605*		1.083*		123*		7.43*		0.548*		6.09*		
Azelaic acid:	C9H16O4															20
Azobenzene:	C12H10N2	1.2														15
B*																
Barium carbonate:	BaCO3	4.43														19
chloride:	BaCl2	3.856														19
chromate:	BaCrO4	4.498														19
fluoride:	BaF2	4.89														19
iodate:	BaI2O6	4.998														19
iodate, hydrate:	BaI2O6.H2O	5.15														19
manganate:	BaMnO4	3.77														19
oxalate:	BaC2O4	2.658														19
oxalate, hydrate:	BaC2O4.H4O2															19
sulfate:	BaSO4	4.5														19
Benzaldehyde:	C7H6O	1.0434	-0.91	1.321	-27	1.622	2.39	153	-180							13,14,4
Benzamide:	C7H7NO	1.341														15
Benzene:	C6H6	0.874	-1	0.61	n=-3.77	1.73	3	145	-300	7.8	25	1.04	3.7	10	100	1,13,4
		0.823*		0.321*		1.88*		131*		9.26*		1.29*		14.8*		
1,3-Benzenediol: (resorcinol)	C6H6O	1.272														4
Benzenethiol:	C6H6S	1.073		1.144		1.575						0.951				14
1,2,3-Benzenetriol:	C6H6O3															4
Benzidine dihydrochloride:	C12H14Cl2N2															
Benzoic acid:	C7H6O2	1.266	-1.8			6.48	22.2									4
3,4-Benzofluoranthene:	C20H12															19
Benzonitrile: (phenyl cyanide)	C7H5N	1.001		1.24	-22.4	1.848										14,4
Benzoperylene:	C22H12															20
Benzophenone:	C13H10O	1.106	-0.86	7.503	n=-6.88	1.485	2.6	139	-120							13,4
1,2-Benzopyrene:	C20H12															15
Benzothiazole:	C7H5NS	1.246														19,15
Benzyl acetate:	C9H10O2	1.0515	-1.3	1.4(45)	n=-5.5	1.63	2.6	146	-219	6.6	23			8.6	53	14,4
Benzyl alcohol:	C7H8O	1.042	-0.78	5.035	-109.8	1.998	5.1	160	-50			1.193	3.4			13,10,4

Benzylamine:	C7H9N	0.981		1.59												15,4,26
Benzylaniline:	C13H13N	1.061		2.31	n=-2.2											15,25
Benzylbenzoate:	C14H12O2	1.112		8.51	-203	1.38	2.9	235	165	3.9	24			6.9	43	14,4
Benzyl n-butyl phthalate:	C19H20O4	1.119														23,15
Benzyl chloride:	C7H7Cl	1.077	-1	1.296	-16.8	1.44	1.84	136	-200	(6.5)	23	(0.994)	2.09	(6.25)	90	13,15,4
Benzyl cyanide:	C8H7N	1.0125		1.98	-38											14
Biphenyl:	C12H10	1.02	-0.8	1.92	n=-5	1.62	3.1	145	-150	(4.93)	21	(1.356)	2.1	27(300)	120	1,13,4
Bis (2-chloroethyl) ether:	C4H8OCl2	1.213	1.17	2.14		1.545										14
Bisphenol A:	C15H16O2															4
Borneol:	C10H180															22,15,23
Bromine:	Br	3.123	-3.36	0.948	-9.56	0.453	0.16	123	-240	15	49	0.226	0.02	5.25	10	13,4
o-Bromoaniline:(2-)	C6H6BrN	1.578		3.19(40)												15,27
m-Bromoaniline:(3-)	C6H6BrN	1.58		5.7	n=-7.3											15,27
p-Bromoaniline:(4-)	C6H6BrN	1.5		1.8(80)												15,27
Bromobenzene:	C6H5Br	1.486	-1.34	1.048	n=-3.8	0.967	0.76	111	-200	13.2(200)	29	.933(200)	1.31	25(200)	140	13,14,4
1-Bromobutane:	C4H9Br	1.2686		0.6073	n=-2.86	0.416	-2.8	104	-190	8.4	29			8.4	50	15,4
2-Bromobutane:	C4H9Br	1.253	ERR	1.434	n=-2.88	1.17	2.5	102	-210	8.8	29			9.8	57	15,4
Bromochloromethane:	CH2ClBr	1.919		0.67												4
Bromodichloromethane:	CHCl2Br	1.98														15
Bromoform:	CHBr3	2.8758	-5.5	1.878	n=-4.1	0.482	-138	98	-138	18.8(200)	40	0.325(200	0.18	9(200)	20	1,13,14,4
Bromonaphthalene:	C10H7Br	1.483	-0.93	4.52	-73											14,4
4-Bromophenyl ether:	C12H8OBr2	1.421														20
1-Bromopropane:	C3H7Br	1.36		0.51	n=-3											15,3
2-Bromopropane:	C3H7Br	1.32		0.46	-21											15,3
o-Bromostyrene:	C8H7Br	1.403														15
o-Bromotoluene:(2-)	C7H7Br	1.423		1.3	n=-0.28											15
m-Bromotoluene:(3-)	C7H7Br	1.01														15
p-Bromotoluene:(4-)	C7H7Br	1.395														23,15
Bromotrichloromethane:	CCl3Br	2.012														4
Bromotrifluoromethane:	CF3Br	1.58		0.15												15
Bromouracil:	C4H3O2N2Br															20
1,2-Butadiene:	C4H6	0.648	-1.12	0.17	-2	2.28	2.89	127	-520	8	25.3	1.482	3.4	14	107	13,4
		0.651*		0.200*		2.20*		126*		7.4*		1.48*		12.5*		
1,3-Butadiene:	C4H6	0.614	-1.12	0.141	-1.6	2.29	3.6	119	-560	7.4	28	1.474	3.73	16	103	1,13,4
		0.650*		0.200*		2.14*		126*		7.57*		1.42*		9.54*		
Butadiene dioxide:	C4H6O2	1.106														19
Butadiene sulfone:	C4H6O2S			0.4(80)		1.809										5,15
Butane:	C4H10	0.579	-1.2	0.133	-3	2.344	3	108	-280	7.7	25	1.68	4.32	15.4	129	1,13,4
		0.603*		0.206*		2.34*		114.7*		7.35*		1.67*		13.7*		
1,2-Butanediol:	C4H10O2	1.0023														20
1,3-Butanediol:	C4H10O2	1.005		98.3	-2753											14,4
1,4-Butanediol:	C4H10O2	1.017		73	n=-11.5	2.24	8.7	210	-47	6.1	24			9.6	72	15,4
2,3-Butanediol: mixture	C4H10O2	1.002														26
1-Butanol:	C4H10O	0.81	-0.7	2.61	n=-7.2	1.34	4	153	-150	(7.55)	22	(1.58)	2.99	(7.25)	130	1,13,4
		0.712*		0.404*		3.20*		127*		9.29*		1.87*		21.7*		
2-Butanol:	C4H10O	0.8026		3.7	-42	2.81										14,4
tert-Butanol:	C4H10O	0.75(50)		4.67	-12	3.18(50)		106(50)		8.18	27	1.567	3.54	12.3	130	13,14,4
		0.710*		0.531*		2.90*		109*		9.4*		1.81*		17.9*		
1-Butene: (a-Butylene)	C4H8	0.595	-1.2	0.075	-3	2.262	3.2	111	-480	7.75	26	1.53	4.1	15	114	1,13,15,4
cis-2-Butene:	C4H8	0.621	-1.1			2.23	3.5			7.59	25	1.41	4	7.6	26	1,15,4
trans-2-Butene: (b-Butylene)	C4H8	0.604	-1.1			2.28	2.8			7.59	25					1,15,4
n-Butyl acetate:	C6H12O2	0.8764	-1.02	0.688	n=-4.2	1.71	2	135	-300	6.5	27	1.48	2.2	9.2	100	2,14,4
2-Butyl acetate:	C6H12O2	0.872	-1.1	0.673	-20											14,4
Butyl acetoacetate:	C8H14O3	0.954														23,15
Butyl acrylate:	C7H12O2	0.894		0.888	n=-4.64	2.02	2.9	136	-258	6.1	22			10.2	57	4
n-Butylamine: (1-Butanamine)	C4H11N	0.741	-1	0.48	n=-4	2.3	1.77	165	-240	6.8	24	1.69	4	9.2	140	2,4
Butylaniline:	C10H15N	0.953														15
n-Butyl benzene:	C10H14	0.86	-0.7	0.95	-15	1.784	3.38	127	-190	(6)	18	(1.452)	3	(7.25)	90	13,4
Butyl benzoate:	C11H14O2	1.01														15,23
n-Butyl cyclohexane:	C10H20	0.8075	-0.7	1.206	-21.6	1.915	3.64	108	-100							13,4
n-Butylcyclohexylamine:	C10H21N	0.839														20
n-Butylcyclopentane:	C9H18	0.785	-0.74	0.828	-12	1.75	3.52	124	-200	(5.7)	16	(1.57)	3.73	(2.75)	110	13
1,2-Butylene oxide:	C2H8	0.837		0.4		2.05		153								23,15

continued

Table Db-1 Selected Physical and Termal Properties of Gases and Liquids for Sizing Equipment (at 25°C unless noted)

	Formula	LIQUID								GAS						
		dens Mg/m	dD/dT x1000	vis mPa.s	dv/dT x1000	cp kJ/kg.K	dc/dT x1000	k mW/m.K	dk/dT x1000	vis µPa.s	dv/dT x1000	cp kJ/kg.K	dc/dT x1000	k mW/mK	dk/dt x1000	References
n-Butylethanolamine:	C6H15NO	0.888		17.4												15
Butyl ether:	C8H18O	0.764	-1	0.83	n=-4	1.67	3.85	117	-100	5.5	23	1.55	3.52	10	120	2, 14
Butyl formate:	C5H10O2	0.891	-1.07	0.666	n=-4.4											14,4
tert-Butyl hydroperoxide:	C4H10O2	0.901														23
n-Butyl lactate:	C7H14O3	0.984														23
Butyl mercaptan:	C4H10S	0.8367		0.473	-4.55	1.915						1.27				14
2-Butyloctanol:	C12H26O	0.831														23
Butyl phenol:	C10H14O	0.908		174	n=-14.3	1.813	3.9	157		5.4	20			10	61	4
Butyl propionate:	C7H14O2	0.8818		0.796	n=-3.8	1.01	-4.05	137		5.9	21			11	66	4
Butyl salicylate:	C11H14O3	1.069														19
Butyl stearate:	C22H44O2	0.8551		8.26												15,13,22
o-n-Butyltoluene:	C11H16	0.865														15
m-n-Butyltoluene:	C11H16	0.853														15
p-o-Butyltoluene:	C11H16	0.851														15
Butyl vinyl ether:	C6H12O	0.7727		0.5(20)	n=-3.14	2.32	2.4	130	-295	6.87	22			11	87	14,15
1-Butyne: (Ethylacetylene):	C4H6	0.652	-1.04	0.16	-3.2	2.488	3.6	113.5	-780	7.7	24	1.507	3.18	13	109	13,4
2-Butyne:	C4H6	0.685	-1.2	0.217	-2.7	2.324	2.7	129	-755	7.28	27	1.451	2.93	12	120	13,15,4
Butyraldehyde: (butanal)	C4H8O	0.796	-1.7	0.38	n=-3.3	2.8	-21	142	-60	7	27	1.6	2.6	12.6	40	2,14,4
n-Butyric acid:	C4H8O2	0.953	-1	1.411	n=-4.6	2.03	4.18	149	-180	(7.3)	23	(1.357)	2.4	(7.5)	100	13,2,4
Butyric anhydride:	C8H14O3	0.962		1.486	n=-4.2	1.8	3.1	137	-228	8.9	20			8	46	14,4
Butyrolactone: (gamma)	C4H6O2	1.122	-0.969	1.7		1.67		276				1.82				14,4
Butyronitrile	C4H7N	0.786	-0.96	0.544	-7.4	2.15	2	153	-220	(7)	23	(1.472)	2.36	(5.5)	100	13,2,4
C*																
Calcium carbonate:	CaCO3	2.71														19
chloride:	CaCl2	2.15														19
fluoride:	CaF	3.18														19
formate:	CaC2H2O4	2.015														19
nitrate:	CaN2O6	2.504														19
oxalate:	CaC2O4	2.2														19
sulfate:	CaO4S	2.96														19
Camphor:	C10H16O															4
Capric acid: (Decanoic acid)	C10H20O2	0.895		4.3(50)												5
e-Caprolactam:	C6H11NO					1.383										14,4
e-Caprolactone:	C6H10O2	1.071														4
Carbon dioxide:	CO2	g				5.96	0.25	86.2	-1163	14.8	47	0.841	0.9	15.95	76	13,4
Carbon disulfide	CS2	1.263	-1.5	0.355	-2.6	1	0.17	168	-343	10.1	34	0.613	0.37	8	40	13,4
Carbon monoxide:	CO	g								17.6	41	1.041	0.15	25	68	13,4
Carbon tetrabromide:	CBr4	3.1455	-2.54	0.4125	-2.26	0.41	0.56	92	-120	11.9	37	0.286	0.08	6.25	10	13
Carbon tetrachloride	CCl4	1.594	-1.9	0.92	-9.6	0.862	0.47	103.1	-223	10.1	31			6.9	32	1,13,4
		1.484*		0.494*		0.92*		92*		11.9*		0.58*		8.6*		
Carbon tetrafluoride:	CF4	g								17	46	0.694	1.25	17	80	13,4
Carbonyl sulfide:	COS	1.036	-1.92	0.01	-3.58	1.265	0.6	121	-1040	12.3	22	0.691	0.71	12	60	13,4
Carene:	C10H16	0.862														10
Cellulose:	(C6H10O5)x	1.3								1.34						
Cellulose acetate:		1.3														
Cellulose triacetate:	C12H16O8									1.32						
Cetyl alcohol:	C16H34O	0.812(60)		13.4(50)												15
Chloral:	C2HCl3O	1.5	-1.8	1.2	-13.7	1.02	0	130				0.6279	0.2	5	10	4
Chlordane:	C10H6Cl8	1.57														16
Chlorine:	Cl2	1.563*		0.496*		0.949*		163*		10.9*		0.497*		6.9*	35	13,1,4
				0.322	n=-1.9					13.4	45	0.477	0.21	9	40	
Chlorine dioxide:	ClO2	3.09														16
Chloroacetic acid:	C3H3ClO2	1.404														20
Chloroacetyl chloride:	C2H2Cl2O	1.418														10,4
Chlorobenzene:	C6H5Cl	1.106	-1.1	0.745	n=-3.4	1.306	1.68	128	-175	7.725	25	0.961	1.85	5	120	13,14,1,4

1-Chlorobutane:	C4H9Cl	0.88	-1.1	0.419	-4.24	1.751	2.2	117	-220			1.163				13,14,4
2-Chlorobutane:	C4H9Cl	0.876		0.4												14,4
Chloro-m-cresol:	C7H7ClO															15
Chlorodifluoromethane:	CHClF2	1.2475	-2.54	0.36(-50)	-7.25	1.141	0.64	87.5	-500	12.8	41	0.647	1.12	11	50	13,4
		1.413*		0.332*		1.1*		119*		10.1*		0.599*		7.15*		
2-Chloroethyl acetate:	C4H7ClO2	1.14														23
2-Chloroethyl ethyl ether:	C4H9ClO	0.989														20
2-Chloroethyl vinyl ether:	C4H7Cl	1.048														20
Chloroform:	CHCl3	1.483	-1.85	0.524	-7.76	0.996	0.75	126	-410	7	40	0.562	0.59	10.5	31	13,4
		1.415*		0.40*		1.0*		111*		11.2*		0.6*		8.75*		
Chloromethyl methyl ether:	C2H5OCl	1.06														10
1-Chloronaphthalene:	C10H7Cl	1.192	-0.77	2.94												23,14
2-Chlorophenol:(o-)	C6H5ClO	1.257		4.1	n=-8.8											15,4
3-Chlorophenol: (m-)	C6H5ClO	1.245		11.55(25)												4,15
4-Chlorophenol:(p-)	C6H5ClO	1.224		4.99(50)												4,15
Chloroprene:	C4H5Cl	0.958	-1	0.38	n=-5	1.36	2	108	-120	7.4	30	0.83	2.4	10.4	103	7,4
3-Chloropropanol:	C3H7Cl	1.131														23,15
o-Chlorostyrene:	C8H7Cl	1.093														23
p-Chlorostyrene:	C8H7Cl	1.08														23
Chlorosulfonic acid:	ClHO3S			2.82	n=-4.8											27
o-Chlorotoluene:	C7H7Cl			1.015	n=-3.6											4
m-Chlorotoluene:	C7H7Cl	1.067	-1.01	0.838	n=-3.45	1.401	2.3	125	-200							15
p-Chlorotoluene:	C7H7Cl			0.87	n=-3.45											4,15
Chlorotrifluoromethane:	CClF3	1.521*		0.318*		1.11*		99.3*		9.83*		0.528*		6.4*		13,4
										14.75	42	0.658	0.78	12.5	60	
5-Chlorouracil:	C3H3ClN2O2															20
Citric acid:	C6H8O7															4
Courmarin:	C9H6O2	0.935														15,20
Creatinine:	C4H7N3O															20
m-Cresol:	C7H8O	1.034	-0.833	14.79	n=-19	2.03	4.9	150	-66	6.45	22	1.39	2.1	8.25	90	13,4
o-Cresol:	C7H8O	1.04	-0.82	4	-40	2.16	1.6	158	-200	6.45	22	1.29	2.5	8.25	100	13,14,4
p-Cresol:	C7H8	1.03	-0.76	6.33	-67	2.089	3	144	-120	6.45	22	1.393	2.1	8.25	90	13,14,4
Crotonaldehyde:	C4H6O	0.849														10,14,4
Cumene: (isopropyl benzene)	C9H12	0.8575	-0.8	0.737	-10.6	1.78	3.3	124.8	-157	6.475	19	1.395	3.06	6.5	100	1,13,14,4
Cyanamide:	CH2N2	1.282														15
Cyanogen:	C2N2	0.9537(-21)								11	34.2	1.097	1	15	70	13,4
Cyanogen chloride:	CNCl									13.1	45.3	0.729	0.6	13	66	13
Cyanogen fluoride:	CNF									19.5	55	0.965	0.58	28	80	13
Cyclobutane:	C4H8	0.7015	-0.9	0.165	-3.8	1.919	3.1	114	-540	8.2	25	1.287	4.9	16	107	13,4
Cyclodecane:	C10H20	0.871														15
Cyclodextrin:																
Cycloheptane: (Suberane)	C7H14	0.811		1.64												4
Cyclohexane: (Hexamethylene)	C6H12	0.7739	-0.97	0.925	-9	1.853	4.4	123	-214	7.12	22.5	1.264	5.1	9	120	13,1,4
		0.715*		0.40*		1.84*		102*		8.6*		1.60*		16.5*		
Cyclohexanol:	C6H12O	0.951	-1.02	16.78	-199	1.885	5.8	136	-160	6.9	24	1.394	3.6	8	120	13,4
Cyclohexanone:	C6H10O	0.94	-0.89	1.9		2.03										14,4
Cyclohexene:	C6H10	0.806	-0.96	0.614	n=-2.4	1.812	4.04	133	-300	7.3	21	1.437	3.76	9.8	110	1,13,4
Cyclohexyl amine:	C6H13N	0.86		1.5	-54											14,4,15
Cyclohexyl chloride:	C6H11Cl	1														23,15
1,3-Cyclooctadiene (cis,cis):	C8H12	0.869														2
1,5-Cyclooctadiene (cis,cis):	C8H12	0.882														15
Cyclopentane:	C5H10	0.739	-0.97	0.415	-4.6	1.81	0.004	131	-328	7.57	22	1.183	4.4	3.75	150	1,13,4
		0.706*		0.32*		1.92*		125*		7.9*		1.34*		16.3*		
Cyclopentanol:	C5H10O	0.976														15
Cyclopentanone:	C5H8O	0.944														4,15
Cyclopentene:	C5H8	0.765	-1.09	0.318	-5.8	1.8	3.5	144	-285	7.8	23	1.103	3.9	8.75	110	13,1,4
Cyclopropane: (Trimethylene)	C3H6	0.623	-0.94	0.30(-50)	-4	1.99	1.7	82	-840	8.7	2.8	1.331	4.9	19	100	13,4
p-Cymene:	C10H14	0.8533		1.8												14,4
Cytosine:	C4H5N3O															20

continued

Table Db-1 Selected Physical and Termal Properties of Gases and Liquids for Sizing Equipment (at 25°C unless noted)

	Formula	LIQUID dens Mg/m	dD/dT x1000	vis mPa.s	dv/dT x1000	cp kJ/kg.K	dc/dT x1000	k mW/m.K	dk/dT x1000	GAS vis µPa.s	dv/dT x1000	cp kJ/kg.K	dc/dT x1000	k mW/mK	dk/dt x1000	References
D*																
2,4-D	C8H6Cl2O3															20,15
DDE	C14H8Cl4															20
DDT	C14H9Cl5															20,15
cis-Decahydronaphthalene:	C10H18	0.8929	-0.76	3	-65	1.68						1.028				4,10,14
trans-Decahydronaphthalene:	C10H18	0.866	-0.75	1.95	-35	1.657						1.215				4,14
Decane:	C10H22	0.73	-0.8	0.921	-9.28	2.173	3.5	126	-280	5.4	16	1.83	3.5	3.75	110	4,13
		0.621*		0.205*		2.79*		91*		8.1*		2.35*		21.8*		
1-Decanol:(Decyl alcohol)	C10H22O	0.83														4,10,15,23
2-Decanol:	C10H22O	0.821														
1-Decene:	C10H20	0.741	-0.777	0.756		2.15						1.6				4,10,13,23
n-Decylbenzene:	C16H26	0.85														4,10
Deuterium oxide:	D2O	1.105				1.714										4
Dextran:	(C6H10O5)x	1.038														3
Diacetone alcohol:	C6H12O2	0.934		1.84	-211	1.88										4,10,14
Diallylamine:	C6H11N	0.783														15
Diallyl ether:	C6H10O															23
Diallyl phthalate:	C14H14O4	1.121														23,20
3,3-Diaminodipropylamine:	C6H17N3	0.925														20
1,3-Diaminopropane:	C3H10N2	0.882														23
Diamyl phthalate:	C18H26O4	1														10,23
1,2,3,4-Dibenzoanthracene:	C22H14															20
Dibenzyl ether:	C14H14O	1.043		4.602	n=-8.6											14,4
o-Dibromobenzene:	C6H4Br2	1.984														20
m-Dibromobenzene:	C6H4Br2															20
p-Dibromobenzene:	C6H4Br2	0.964(100)														15
Dibromochloromethane:	CHClBr2	1.546														15
Dibromochloropropane:	C3H5ClBr2															23
Dibromomethane:	CH2Br2	2.4956														4,15
Dibutylamine:	C8H19N	0.7577	-1	0.89	n=-4	1.96	1.77	133	-240	5.2	24	1.59	4	10.5	140	2,14,4
N,N-Dibutylethanolamine:	C10H23NO	0.856														23,15
Dibutylisopropanolamine:	C11H25NO	0.837														23
Dibutyl maleate:	C12H20O4	0.9907		4.76	-175											14,4
Di-n-butyl phthalate:	C16H22O4	1.042	-0.82	16.47		1.79										14,4
Di-n-butyl sebacate:	C18H34O4	0.9324		7.96	-214											14,15,4
Dichloroacetyl chloride:	C2HCl3O	1.5315														4,15
o-Dichlorobenzene:	C6H4Cl2	1.305		1.324	-17.7	1.1	1.9	121	-164	7.3	29			7.2	48	14,4
m-Dichlorobenzene:	C6H4Cl2	1.2828		1.026	-9.7	1.1	2	117	-152	7.3	29			7.4	49	14,4
p-Dichlorobenzene:	C6H4Cl2	1.248	-1.17	1.06	-7.15	1.16	1.6	112	-168	7.5	29			7.4	49	14,4
3,3-Dichlorobenzidine:	C12H10N2Cl2															
Dichlorodifluoromethane:	CCl2F2	1.311	-2.84	0.26	-1.96	0.972	0.5	71	-430	12.3	27.5	0.607	0.87	9.6	54	13,4
		1.486*	-3.37*	.373*	2.4*	.896*	1.5*	95.1*	-470*	10.3*	50*	0.569*	2.5*	6.9*	*50	
1,1-Dichloroethane:	C2H4Cl2	1.168	-1.62	0.453	-5.34	1.28	1.26	113	-220	9.375	31	0.788	1.3	9.5	60	13,14,4
1,1-Dichloroethylene:	C2H2Cl2	1.213	-1.67	0.4175	-9.9	1.18	1.52	107	-271	10.325	33	0.727	0.92	8.5	60	13,4
cis-1,2-Dichloroethylene:	C2H2Cl2	1.2757	-1.6	0.444		1.18										14,4
trans-1,2-Dichloroethylene:	C2H2Cl2	1.2463	-1.68	0.385	-3.8	0.69										14,4
Dichlorofluoromethane:	CHCl2F	1.366		0.263	-3.4	1.072		100	-370	11.3	30	0.586		8.5	52	13,4
		1.406*		0.366*		1.04*		114*		11*		0.721*		7.9*		
Dichloroisopropyl ether:	C6H12Cl2O	1.106														23,22
Di (2-chloroisopropyl) ether:	C6H12Cl2O	1.11														22,23
Dichloromethane:	CH2Cl2	1.323	-1.629	0.389	-10.5	1.16	0.43	139	-315	10.23	33	0.622	0.92	8.25	50	13,14,4
1-Dichloro-2,4,-nitrobenzene:	C6H13Cl2NO2	1.44(80)														15
1,5-Dichloropentane:	C5H10Cl2	1.1														4,15
2,3-Dichlorophenol:	C6H4Cl2O															15
2,4-Dichlorophenol:	C6H4Cl2O															15
2,5-Dichlorophenol:	C6H4Cl2O															15

2,6-Dichlorophenol:	C6H4Cl2O															15
3,4-Dichlorophenol:	C6H4CL2O															19,20
1,3-Dichloropropane:	C3H6Cl2	1.187														15
1,3-Dichloro-2-propanol:	C3H6Cl2O	1.351														15
1,3-Dichloropropene:	C3H4Cl2	1.217														15
Dichlorotetrafluoroethane:	C2Cl2F4	1.456	-2.8	0.368	-12	1.02	0.68	68.6	-310	11.3	13.3	0.667	1.07	11.2	53	13,4
2,4-Dichlorotoluene:	C7H6Cl2	1.246														4,15
Dicyandiamide:	C2H4N4	1.40														20
cis-Dicyano-1-butene:	C6H6N2															4
trans-Dicyano-1-butene:	C6H6N2															4
1,4-Dicyano-1-butene:	C6H6N2															4,20
Dicyclohexylamine:	C12H23N	1.482														4,15,23
Dicyclopentadiene:	C10H12	0.93														4,15,23
Dieldrin:	C12H8Cl6O	1.75														20
Diesel fuel:		0.875	-0.67					116.3	-117							23
Diethanolamine:	C4H11NO2	1.087	-0.64	471	-18360	2.22										14,4
1,1-Diethoxybutane:	C8H18O2	0.841														19
2,5-Diethoxytetrahydrofuran:	C8H20O3	0.962														20
Diethylamine:	C4H11N	0.712	-1.06	0.31	-4	2.43	3.5	133.6	-280	7.3	25	1.62	3.78	13	120	4,13,10
Diethylamine hydrochloride:	C6H12ClN	1.048														19,20
3-(Diethylamino)propylamine:	C7H18N2	0.826														23
N,N-Diethylaniline:	C10H15N	0.931	-0.8	2.02	n=-8	1.83	2.5	136	-110							13,4
2.6-Diethylaniline:	C10H15N	0.906														4
1,2-Diethylbenzene:	C10H14	0.88		1.13	n=-4.5	1.8	3.8	126	-195	6.1	17			10.8	69	4
1,3-Diethylbenzene:	C10H14	0.864		0.995	n=-3.8	1.8	3.7	123	-190	5.7	19			9.7	70	4
1,4-Diethylbenzene:	C10H14	0.862		1.06	n=-3.8	1.8	3.9	123	-196	6.1	16			10.6	68	4
Diethyl carbonate:	C5H10O3	0.969	-1.1	0.748		1.79										14,4
Diethyl disulfide:	C4H10S2	0.998														4
Diethylene glycol:	C4H10O3	1.118	-0.75	34	n=-11.5	2.31										14,4
-butyl ether: (mono-n-)	C8H18O3	0.948		6.25		2.29										23,11,4
-butyl ether acetate:	C10H20O4	0.98		3.35												23,1
-diacetate:	C8H14O5	1.1														23
-dibutyl ether:	C12H26O3	0.9														23,11
-diethyl ether:	C8H18O3	0.9														23,11
-dimethyl ether:(1,2-)	C6H14O3	0.945		1.75												11,4
-ethylbutyl ether:	C10H22O3															11
-ethyl ether: (mono)	C6H14O3	0.9841		3.71	-53	2.25										14,11,4
-ethyl ether acetate:	C8H16O4	1.0096	-1.02	2.8		2.26										14,11,4
-ethylhexyl ether: (2-)	C12H26O3															11
-hexyl ether: (n-)	C10H22O3	0.934		8.35												11
-methyl ether: (mono)	C5H12O3	1.0167		3.48	-53	2.26										14,11,4
-methyl ether acetate:(2-)	C7H14O4	1.04														23,11
-phenyl ether:	C10H14O3	1.115		46.05												11
-phthalate:	C16H22O6	1.1														23
-vinyl ethyl ether:	C8H16O3	0.939		1.35												11
Diethylenetriamine:	C4H13N3	0.959		4.48	n=-6.3	1.82	4.22	141	-217	6.43	21			10.6	74	4
N,N-Diethylethanolamine:	C6H15NO	0.884														23,15
Diethyl fumarate:	C8H12O4	1.046														23,15
Di(2-ethylhexyl)amine:	C16H35N	0.801														19,22,23
Diethyl ketone:	C5H10O	0.809	-0.94	0.442	-3.7	2.23	2.47	141	-240	(6.73)	21	(1.554)	2.73	(6)	120	13,14,4
Diethyl maleate:	C8H12O4	1.063		3.14												14
Diethyl oxalate:	C6H10O4	1.075		1.99	n=-5											4,15
3,3-Diethylpentane:	C9H20	0.749														23
Diethyl phthalate:	C12H14O4	1.114														4,23
Diethyl pimelate:	C11H20O4	0.988														19
Diethyl siloxane: (Silicone)		0.952	-0.7	8.8	-178	2.07	1.1	132	-210							1
Diethyl succinate:	C8H14O4	1.035														4,23,15
Diethyl sulfate:	C4H10O4S	1.177														4,23,15
Diethyl sulfide:	C4H10S	0.832	-0.98	0.422	-4.8	1.89	1.04	135	-320							13,14,4
Difluoromethane:	CH2F2									12.3	38	0.824	1.47	14	70	13,4
Diglycol chlorohydrin:	C4H9ClO2	1.168														23
Dihexylamine:	C12H27N	0.795														15,23

continued

Table Db-1 Selected Physical and Termal Properties of Gases and Liquids for Sizing Equipment (at 25°C unless noted)

	Formula	LIQUID								GAS						
		dens Mg/m	dD/dT x1000	vis mPa.s	dv/dT x1000	cp kJ/kg.K	dc/dT x1000	k mW/m.K	dk/dT x1000	vis µPa.s	dv/dT x1000	cp kJ/kg.K	dc/dT x1000	k mW/mK	dk/dt x1000	References
Dihexyl ether:	C12H26O	0.789														4,15
Diisobutylene:	C8H16	0.711														15
Diisobutyl ketone:	C9H18O	0.802														15,23
Diisopropylamine:	C6H15N	0.71	-1	0.69	n=-4	2.05	1.77	113	-240	6.8	24	1.59	4	10.5	140	2,4
N,N-Diisopropylethanolamine:	C8H19NO	0.87														23
Diisopropyl maleate:	C10H16O4	1.005														23
Diketene:	C4H4O2	1.073														15,23
1,1-Dimethoxyethane:	C4H10O2	0.852														15
2,2-Dimethoxypropane:	C5H1202	0.847														
N,N-Dimethylacetamide:	C4H9NO	0.937	-0.86	1.79	-68	2.02										14,4
Dimethylamine:	C2H7N	0.651	-1.12	0.132	-4.6	3.07	1.75	141	-400	9.1	28	1.53	4.08	16	93	13,4
4-Dimethylaminoazobenzene:	C14H15N3															20
3-(Dimethylamino)propionitrile:	C5H10N2	0.866														15,23
3-(Dimethylamino)propylamine:	C5H14N2	0.812														23
N,N-Dimethylaniline:	C8H11N	0.951	-0.81	1.33	n=-4.7	1.81	2.9	142	-175							15,23
2,2-Dimethylbutane:	C6H14	0.6444	-0.52	0.359	-3.36	2.19	4.6	104	-34	6.6	23.4	1.62	5.44	12	120	13,14,4
2,3-Dimethylbutane:	C6H14	0.656	-0.92	0.363	-4.38	2.19	5.46	106	-280	7	20	1.71	4.27	10.25	130	13,14,4
1,3-Dimethyl-1-butanol:	C6H14O	0.808														15,23
2,2-Dimethylbutanol:	C6H14O	0.824														15
2,3-Dimethylbutanol:	C6H14O	0.826														15
2,4-Dimethylbutanol:	C6H14O	0.809														15
2,3-Dimethyl-2-butanol:	C6H14O	0.819														15
2,3-Dimethyl-1-butene:	C6H12	0.672														19,23,4
2,3-Dimethyl-2-butene	C6H12	0.708														15,23,4
1,3-Dimethylbutyl acetate:	C8H16O2	0.863														22,23
1,3-Dimethylbutylamine:	C6H15N	0.743														22,23
1,1-Dimethylcyclohexane:	C8H16	0.775														4
cis-1,2-Dimethylcyclohexane:	C8H16	0.769														4,15
trans-1,2-Dimethylcyclohexane:	C8H16	0.7772														4,15
1,1-Dimethylcyclopentane:	C7H14	0.749	-0.92									1.36	4.3			4,1
2,6-Dimethyl-1,4-dioxane:	C6H12O2	0.939														23
Dimethyl disulfide:	C2H6S2	1.046														4,15,22
N,N-Dimethylethanolamine:	C4H11NO	0.882														23,15
Dimethyl ether:	C2H6O	0.661	-1.48	0.09	-2.8	2.27	1.06	144	-666	9.31	32	1.42	2.9	17	100	13,4
N,N- Dimethylformamide:	C3H7NO	0.944	-1.07	0.802	-7.3	2.06	2.57	184	-326	6.4	25	1.26	3.3	8.2	73	T,14,4
2,5-Dimethylfuran:	C6H8O	0.903														15,23
2,2-dimethylheptane:	C9H20	0.705														19
2,3-Dimethylheptane:	C9H20	0.726														19
2,4-Dimethylheptane:	C9H20	0.714														19
2,5-Dimethylheptane:	C9H20	0.719														19,22
2,6-Dimethylheptane:	C9H20	0.709														19,22
3,3-Dimethylheptane:	C9H20	0.725														19,22
3,4-Dimethylheptane:	C9H20	0.731														19,22
3,5-Dimethylheptane:	C9H20	0.722														19,22
4,4-Dimethylheptane:	C9H20	0.722														19,22
2,6-Dimethyl-4-heptanol:	C9H20O	0.809														4,20
2,2-Dimethylhexane:	C8H18	0.695														4
2,3-Dimethylhexane:	C8H18	0.712														4
2,4-Dimethylhexane:	C8H18	0.696														4
2,5-Dimethylhexane:	C8H18	0.688														4
3,3-Dimethylhexane:	C8H18	0.705														4
3,4-Dimethylhexane:	C8H18	0.704														4
1,1-Dimethylhydrazine:	C2H8N2	0.7869	-1			2.73	4.27					1.05	0			10,15,23
Dimethylisopropanolamine:	C5H13NO	0.845														23
Dimethyl maleate:	C6H8O4	1.146		3.21		1.8										14
2,6-Dimethylmorpholine:	C6H13NO	0.928														15,23
Dimethylnitrosoamine:	C2H6NO	1.005														19

2,2-Dimethyloctane:	C10H22	0.719													23
2,3-Dimethyloctane:	C10H22	0.732													19
2,4-Dimethyloctane:	C10H22	0.721													22
2,5-Dimethyloctane:	C10H22	0.731													22
2,6-Dimethyloctane:	C10H22	0.723													22
2,7-Dimethyloctane:	C10H22	0.719													22
3,3-Dimethyloctane:	C10H22	0.734													22
3,4-Dimethyloctane:	C10H22	0.741													22
3,5-Dimethyloctane:	C10H22	0.731													22
3,6-Dimethyloctane:	C10H22	0.731													22
4,4-Dimethyloctane:	C10H22	0.732													22
4,5-Dimethyloctane:	C10H22	0.742													22
2,3-Dimethylpentaldehyde:	C7H14O	0.823													15,23
2,2-Dimethylpentane:	C7H16	0.674													4
2,3-Dimethylpentane:	C7H16	0.6909		0.42	-1.6	2.14						1.66				14
2,4-Dimethylpentane:	C7H16	0.668		0.345	-3	2.139						1.66				14
3,3-Dimethylpentane:	C7H16	0.694														22,20
2,3-Dimethyl-3-pentanol:	C7H16O	0.883														19
2,3-Dimethylphenol:	C8H10O															15
2,4-Dimethylphenol:	C8H10O	1.027														15
2,5-Dimethylphenol:	C8H10O	0.965(80)														15
2,6-Dimethylphenol:	C8H10O															15
3,4-Dimethylphenol:	C8H10O	1.064														15
3,5-Dimethylphenol:	C8H10O	1.008														15
Dimethyl o-phthalate:	C10H10O4	1.19														4,23,15
Dimethyl m-phthalate:	C10H10O4	1.194														15
Dimethylpimelate:	C9H16O4	1.034														15
2,2-Dimethylpropane:	C5H12	0.581	-1	0.228	-2.36	2.22	7	86	-266	7.1	21	1.687	4.52	15	130	13, 14
		0.603*		0.28*		2.14*		90*		6.89*		1.69*		13.2*		
2,2-Dimethylpropanol:	C5H12O															23,4
Dimethyl sulfide:	C2H6S	0.8423	-1.16	0.279	-4.23	1.9	1.2	124	-357	8.63	25	1.2	2.13	13.75	110	13,14,4
Dimethyl sulfone:	C2H6O2S	1.17														15
Dimethylsulfoxide:	C2H6OS	1.0958	-0.977	1.996	-34	1.95	1.7					1.14				14, 10
Dimethyl terephthalate:	C10H10O4															4,20,15
2,5-Dimethyltetrahydrofuran:	C6H12O	0.833														15
o-,1,2-Dinitrobenzene:	C6H4N2O4	1.565														19
m-Dinitrobenzene:(1,3-)	C6H4N2O4	1.575														19
p-Dinitrobenzene:(1,4-)	C6H4N2O4	1.625														19
4,6-Dinitro-o-cresol:	C7H10N2O5															15
2,4-Dinitrophenol:	C6H4N2O3	1.683														15
2,4-Dinitrotoluene:	C7H6N2O4	1.32(70)														15
Dioctyl phthalate:	C24H38O4	0.974	-1	4.7	-415	2	0	134	-96							10
Dioctyl sulfosuccinate-sodium salt:	C20H37NaO7S															20
1,3-Dioxane:	C4H8O2															20
1,3-Dioxolane	C3H6O2	1.06														20
Dipentene:	C10H16	0.836	-1	0.79	26.7	1.867	3.5	151	0							10,23
Diphenylamine:	C12H11N	1.16														15,23
Diphenyl ether:	C12H10O	1.0704	-0.86	2.74	-30	1.58										14,4
1,2-Diphenyl hydrazine:	C12H12N2	1.158														15
Diphenylmethane:	C13H12	1	-0.8	2.23	-17.4	1.566	2.58	134.5	-100							13,4
Dipropylamine:	C6H15N	0.7329	-0.8	0.502	-6.3	2.5	0					1.63	4.14			14,10,4
Dipropylene glycol:	C6H14O3	1.022	-0.73	106		2.428	4.6	160	0			1.51	3.2			10,11,4
Dipropylene glycol (mixed isomers)																23
-butyl ether: (mono n-)	C10H22O3	0.913		1.3												15
-ethyl ether: (mono)	C8H18O3	0.93		1.1												15
-isopropyl ether:	C9H20O3	0.878														15
-methyl ether: (mono)	C7H16O3	0.9608		3.5												23,15
Dipropyl ether:	C6H14O	0.7419	-0.98	0.448	-4.78	2.138	3.84	127	-180	6.16	23.1	1.56	3.52	10	120	13,14,4
Dipropyl ketone:	C7H14O	0.811	-0.84	0.718	-6.7	2.198	2.67	139	-220							13
Dipropyl oxalate:	C8H14O4	1.018		2.73	n=-5.6											19
Dipropyl sulfone:	C6H14O2S	1.028(50)														4,15

continued

Table Db-1 Selected Physical and Termal Properties of Gases and Liquids for Sizing Equipment (at 25°C unless noted)

	Formula	LIQUID								GAS						
		dens Mg/m	dD/dT x1000	vis mPa.s	dv/dT x1000	cp kJ/kg.K	dc/dT x1000	k mW/m.K	dk/dT x1000	vis µPa.s	dv/dT x1000	cp kJ/kg.K	dc/dT x1000	k mW/mK	dk/dt x1000	References
Divinyl acetylene:	C6H6	0.776														23,22
Divinyl benzene:	C10H10	0.912														23,22,4
Divinyl ether:	C4H6O	0.769														4
Docosanoic acid:	C22H44O2	0.822(100)				2.326										25
Dodecane:	C12H26	0.7452	-0.719	1.378	-12.8	2.189	3.27	140	-220	(5.25)	14	1.98	2.89	31(300)	110	13,14,4
Dodecanoic acid:	C12H24O2	0.869														4
1-Dodecanol:	C12H26O	0.831														4
2-Dodecanol:	C12H260	0.825														19
1-Dodecene:	C12H24	0.754	-0.73	1.19	-7.16	1.87	15.6	134.6	-210			1.6	4.3			10,4
Dodecylamine:	C12H27N															4,15
Dodecyl benzene:	C18H30	0.855														4,19
Dowtherm A:		1.056	-0.812	3.625	-35	1.588	2.83	140	-113							13
		0.852*	-1.12*	0.273*	-1.12*	2.24*	3*	112*	-120*	10.1*	22*	1.83*	2.4*	20.4*	92*	
Dowtherm J:		0.876	-0.8	0.848	-9.5	1.85	3.2	132	-82	6	19	1.34	4			13, T
		0.729*		0.172*		2.40*		118.5*		8.6*		1.91*				
E*																
n-Eicosane: (Didecyl)	C20H42	0.789	-0.7	3.895	-24.75	2.2	3.3	150	-180	7.6(400)	17	3.05(400)	4.6	31(400)	180	13,4
Eicosanoic acid:	C20H40O2	0.824(100)														5
a-Endosulfan:	C9H6Cl6O3S															
Endrin:	C12H8Cl6O															
Epichlorohydrin:	C3H5OCl	1.174	-1.27	1.03	-17.6	1.436	2.26	72.5	116	8.3	32	0.678	2.19	9.5	56	10,14,4
3,4-Epoxy-1-butene:	C4H6O	0.87														
Ethane:	C2H6	0.546	-1.5	0.168	-1.13	2.42	8.7	157	-780	9.35	29	1.754	4.18	21	160	13,4
Ethanol:	C2H60	0.789	-0.86	1.166	n=-5.5	2.45	11.4	172	-240	8.73	29	1.608	2.89	11	160	13,14,4
		0.757*		0.4287*		3.00*		153.7*		10.4*		1.83*		19.9*		
95%				1.31	n=-6			207								
40%				2.67	n=-9.5			387								
2-Ethoxy-3,4-dihydro-2-pyran:	C7H12O2	0.963														23,22
3-Ethoxy-1-propanol:	C5H12O2	0.904														23
3-Ethoxypropionaldehyde:	C5H10O2	0.911														23
3-Ethoxypropionic acid:	C5H10O3	1.041														23
Ethyl acetate:	C4H8O2	0.89455	-1.3	n=-2.87	-3	1.9	1.76	149	-300	7.55	26	1.24	2.9	11.1	74	1,13,14,4
Ethyl acetoacetate: enol form	C6H10O3	1.021	-0.97	1.508		1.93										14,4
Ethyl acrylate:	C5H8O2	0.924		0.577	n=-3.8	1.967										14,4
Ethyl amine:	C2H7N	0.68	-1.08	0.19	-5.2	3.07	1.6	185	-260	8	28	1.613	4	15	106	13,4
Ethyl amyl ketone:	C8H16O	0.823														23
N-Ethylaniline:	C8H11N	0.963		2.04	-17											4
Ethylbenzene:	C8H10	0.862	-0.88	0.6373	-5.9	1.762	5.44	130	-270	6.7	20	(1.881)	1	4.5	100	13,14,4
Ethylbenzoate:	C9H10O2	1.041	-0.9	1.98	-17	1.615	3	141	-190							13,14,4
Ethyl bromide (Bromomethane):	C2H5Br	1.45		0.379	-21	0.926		121								14,4
2-Ethyl-1-butanol:	C6H14O	0.8295	-0.78	5.892												14,4
2-Ethyl-1-butene:	C6H12	0.689														4
2-Ethylbutyl acetate:	C8H16O2	0.874														23
N-Ethylbutylamine:	C6H15N	0.776														23
Ethylbutyl ether:	C6H14O	0.745														23
Ethylbutyl ketone:	C7H14O	0.818														23
2-Ethylbutyraldehyde:	C6H12O	0.8162														23,15
Ethylbutyrate:	C6H12O2	0.8739	-1.06	0.6127	-5.05	1.901	2.5	136	-220							13,14,4,23
2-Ethylbutyric acid:	C6H12O2	0.924														4,23
Ethyl chloride:	C2H5Cl	0.8905	-1.34	0.2345	n=-2.6	1.6	1.42	120	-300	9.75	31	0.971	2.1	11	93	13,14,4
Ethylchloroacetate:	C4H7ClO2	1.15				1.74										15,23
Ethyl chloroformate:	C3H5ClO2	1.135		0.547	n=-3.6	1.81	4.4	152		8.1	27			10.1	56	4
Ethyl cinnamate:	C11H12O2	1.049		8.7												14
Ethyl crotonate:	C6H10O2	0.918														
Ethylcyclohexane:	C8H16	0.7839	-0.81	0.787	-7.7	1.88	4.41	110	-140	6.05	18	1.562	4	5	120	13,14,4

N-Ethylcyclohexylamine:	C8H17N	0.844														
Ethylcyclopentane:	C7H14	0.763	-0.92	0.544	-4.84	1.896	4.5	120	-250	6.8	20	1.499	3.85	7	120	4
N-Ethyldiethanolamine:	C6H15NO2	1.014														
Ethylene:	C2H4									10.2	33	1.553	3.69	21	140	13,4
		0.5679*		0.162*		2.32*		192*		6.04*		1.31*		6.44*		
Ethylene carbonate:	C3H4O3	1.321														14,4
Ethylene chlorohydrin:	C2H5ClO	1.201	-1.1	3.1	-82											14,4
Ethylene cyanohydrin:	C3H5NO	1.041	-0.76													14,4
Ethylenediamine:	C2H8N2	0.8906	-0.937	1.54	-33	2.93										14,4
Ethylene diamine tetra-acetic acid:	C10H16N2O8															
Ethylene dibromide:	C2H4Br2	2.4834	-2.72	0.97	-11	0.606	0.18	107	-180	12.85	42	0.328	0.38	4.25	50	13,14
Ethylene dichloride:	C2H4Cl2	1.235	-1.56	0.78	-11.06	1.27	1.92	134	-240	9.6	32	0.816	1.12	7.75	70	13,14,4
Ethylene glycol:	C2H6O2	1.11	-0.7	1.857	-36.8	2.39	5.86	256	111	8.2	28	1.42	2.3	9.75	110	13,14,4
-benzyl ether:	C9H12O2	1.064														19,22
-butyl ether: (mono-n-)	C6H14O2	0.8894		3.15		2.341										14,4,11
-butyl ether acetate:	C8H16O3	0.936	-1	1.627	-33	1.91	4.2	151								10,11
-butyl methyl ether:	C7H16O2	0.841														11
-diacetate:	C6H10O4	1.128	-1.13	2.74												14,4,11
-dibutyl ether:	C10H22O2	0.8374		1.14												11
-diethyl ether:	C6H14O2	0.8417		0.54												11
-diformate:	C4H6O4	1.227		2.44												11
-dimethyl ether:(1,2-)	C4H10O2	0.8655		0.455												14,4,11
-dipropionate:	C6H14O4	1.0484														11
-ethylbutyl ether: (2-)	C8H18O2	0.833														11
-ethyl ether: (mono)	C4H10O2	0.9252		1.85		2.326										14,4,11
-ethyl ether acetate:	C6H12O3	0.975	-0.79	1.025												14,4,23
-ethyl ether acrylate:	C7H12O3	0.9834		1.45												11
-ethylene ether:	C4H8O2	1.034	-0.86	1.2175	-10.5	1.736	2.99	163	-50	(7.6)	27	(1.16)	3.04	(7.25)	130	13,4,11
-ethylhexyl: (2-)	C10H23O2	0.885														11
-hexyl ether: (n-)	C8H18O2	0.8887		5.04												11
-methylene ether:	C3H6O2	1.06														23,19,11
-methyl ether: (mono)	C3H8O2	0.9602		1.6		2.203										14,11
-methyl ether acetate:(2-)	C5H10O3	1.0049		0.94												14,11
-methyl ether acrylate:	C6H10O3	1.009														11
-methylethyl ether:	C6H10O2	0.8529														19
-phenyl ether:	C8H10O2	1.109														11
-phenyl ether acetate:	C10H12O3	1.108		9.24												11
-vinyl butyl ether:	C8H16O2	0.865		0.88												11
Ethylene imine:	C2H5N	0.832	-1.4	0.418	n=-2.7	2.46	4.6	188	-550	7.6	24	1.15	3.6	13.8	100	14,2,10,4
Ethylene oxide:	C2H4O	0.871	-1.3	0.195	n=-2.9	1.997	1.84	194	-540	9.4	32	1.097	2.9	12	93	4
N-Ethylethanolamine:	C4H11NO	0.913														23
Ethyl ether:	C4H10O	0.708	-1.3	0.221	-4.39	2.35	2.87	137	-300	7.58	24	1.594	3.3	19	80	13,14,4
Ethyl-3-ethoxyacrylate:	C7H12O3	0.998														20
Ethyl formate:	C3H6O2	0.917	-1.3	0.38	-3	1.9	2.8	146	-4180	9	32	1.19	2	11.7	84	1,14,2,4
3-Ethylheptane:	C9H20	0.722														22
4-Ethylheptane:	C9H20	0.725														19,22
2-Ethylhexaldehyde:	C8H16O	0.815														20
3-Ethylhexane:	C8H18	0.708														4
3-Ethylhexanoic acid:	C8H16O2	0.903														20
2-Ethyl-1-hexanol:	C8H18O	0.8291	-0.73	9.8		2.37										14,4
2-Ethyl-1-hexene:	C8H16	0.722														4
2-Ethylhexyl acetate:	C10H20O2	0.867	-0.87	1.5		2.09										14,4
2-Ethylhexyl acrylate:	C11H20O2	0.885		1.57	n=-6.7											4
2-Ethylhexylamine:	C8H19N	0.789														15
2-Ethylhexyl chloride:	C8H17Cl	0.877														
Ethyl iodide:	C2H5I	1.9244		0.56	-21	0.74		110				0.415				14,4
n-Ethyl lactate:	C5H10O3	1.0272	-1.12	2.44		2.15										14,4
Ethyl mercaptan:	C2H6S	0.814	-1.08	0.277	-4.57	1.89	1.2	130	-557	8.48	27	1.189	2.24	11	80	4
Ethyl methacrylate:	C6H10O2	0.917														4
3-Ethyl-2-methylhexane:	C9H20	0.726														19,22
3-Ethyl-3-methylhexane:	C9H20	0.736														19,22

continued

Table Db-1 Selected Physical and Termal Properties of Gases and Liquids for Sizing Equipment (at 25°C unless noted)

	Formula	LIQUID								GAS						
		dens Mg/m	dD/dT x1000	vis mPa.s	dv/dT x1000	cp kJ/kg.K	dc/dT x1000	k mW/m.K	dk/dT x1000	vis µPa.s	dv/dT x1000	cp kJ/kg.K	dc/dT x1000	k mW/mK	dk/dt x1000	References
4-Ethyl-2-methylhexane:	C9H20	0.718														19,22
4-Ethyl-3-methylhexane:	C9H20	0.737														22
3-Ethyl-2-methylpentane:	C8H18	0.714														19,15
3-Ethyl-3-methylpentane:	C8H18	0.722														19,15
N-Ethylmorpholine:	C6H13NO	0.905		0.83												23,15,11
1-Ethylnaphthalene:	C12H12	1.004	-0.72	3.087	-48.6	1.644	2.7	129	-100	(5.625)	17	(1.469)	2.18	(5)	80	13,23
2-Ethylnaphthalene:	C12H12	0.988	-0.73	2.26	-27.8	1.643	2.7	128	-100	(5.95)	18	(1.4425)	2.22	(5)	80	13
3-Ethyloctane:	C10H22	0.735														23
4-Ethyloctane:	C10H22	0.735														23
Ethyl oxalate:	C6H10O4	1.07	-1.16	1.85	-4.6	1.813										14
3-Ethylpentane:	C7H16	0.693														4,15
Ethyl propionate:	C5H10O2	0.874	-1.12	0.431	-5	1.963	3.1	137	-260	7.23	25	1.326	2.87	10.3	90	13,14,4
2-Ethyl-3-propylacrolein:	C8H14O	0.846														23,22
2-Ethyl-3-propylacrylic acid:	C8H14O2	0.943														23,22
Ethyl propyl ether:	C5H12O	0.728	-0.96	0.3083	-3.14	2.236	2.62	127	-320							13,4
Ethyl propyl ketone:	C6H12O	0.815														20
Ethyl salicylate:	C9H10O3	1.213		1.77(45)												15
m-Ethylstyrene:	C10H12	0.889														
Ethyl sulfide:	C4H10S	0.8312		0.426	-21	1.902						1.298				14
m-Ethyltoluene:	C9H12	0.859														4
0-Ethyltoluene:	C9H12	0.875														4
p-Ethyltoluene:	C9H12	0.859														4
Ethyl valerate:	C7H14O2	0.877		0.847												3
Eugenol:	C10H12O2	1.066		8.1	n=-14											15
F*																
Fluoranthene:	C16H10	1.252														4
Fluorene:	C13H10	1.203														4
Fluorine:	F2			0.275(80	-13.9					24	60			26.9	76	13,4
		1.524*		0.237*		1.55*		158*		7.23*		0.795*		7*		
Fluorobenzene:	C6H5F	1.018	-1.24	0.559	-7.8	1.531	1.34	125.6	-280	6.1	36	1.02	2.74	12.5	100	13,14,4
Fluoromethane:	CH3F	1.1951														15
5-Fluorouracil:	C4H3FN2O2															20
Formaldehyde:	CH2O	0.747	-1.48	0.09	-2	2.92	6	208	-840	9.6	32	1.112	1.13	14	70	13,4,16
Formamide:	CH3ON	1.11	-0.4	3.44	-64	2.387	1	360	-333	(7.25)	26	(1.08)	1.9	(5.25)	26	13,4,16
Formamidoxime:	CH4N20															
Formic acid:	CH2O2	1.214	-1.2	1.663	25.3	2.175	1.1	259	-367	(988)	31	(1.117)	1.32	(11.25)	90	13,14,4
Fuel T-5:		0.839	-0.7	2.9	-73	1.972	4.4	116	-118	(1.7)	42	(2.11)	1.6	(6.8)	112	1
Fuel oil: No. 4:		0.915		30				125	-65							3
No. 6:		0.985		5000				116	-65							3
Fumaric acid:	C4H4O4	1.635														
Furan:	C4H4O	0.93	-1.31	0.3656	-3.3	1.68	2.67	152	-333	8.6	28	1	2.89	95	100	13,14,4
Furfural:	C5H4O2	1.155	-1.06	1.503	-25.5	1.657	4.18	171	-220	(7.45)	26	(1.23)	1.61	(7.25)	90	13,14,4
Furfuryl alcohol:	C5H6O2	1.0483	-1.1	4.62		2.010922										14,4
G*																
Gasoline:		0.747	-0.85	0.5035	-5.1	2.08	4.5	115	-199	(2.61)	31	(2.0175)	2.5	(14.4)	98	1,13
Glucose:	C6H12O6	1.562														15
Glycerol:	C3H8O3	1.257	-0.6	875	-120900	2.38	5.5	279	133	(7.6)	24	(1.765)	1.4	(2.5)	100	1,13,14,4
Glyceryl triacetate:	C9H14O6	1.596														4
Guanine:	C5H5N5O															

H*

Helium:	He	0.125*		0.0365*		4.48*		31.2*		1.09*		5.19*		8.3*		4
										19.6	40	5.2	0	150	310	
Heptachlor:	C10H5Cl7	1.57														
Heptadecane:	C17H36	0.776	-0.72	2.71	-22.7	2.198	3.06	144	-200	9(400)	24	3.05(400)	4.6	35(400)	190	13,4
Heptadecanoic acid:	C16H32O2	0.853(60)														5
n-Heptane:	C7H16	0.684	-0.86	0.39	-4.5	2.21	3.26	122	-340	6.18	19	1.85	2.3	8	120	13,4
		0.614*		0.201*		2.57*		98*		7.3*		1.98*		18*		
2,4-Heptanediol:	C7H16O2	0.926														22
Heptanoic acid:	C7H14O2	0.853(60)		3.4	-185											15,4
1-Heptanol:	C7H16O	0.818	-0.88	5.79	-244	2.427	14	136	-140	(5.8)	24	(1.64)	3.13	(7.75)	111	13,4
2-Heptanol:	C7H16O	0.81834		4.66	-43			136	-264	5.83	21					14,4
3-Heptanol:	C7H16O	0.817														
2-Heptanone:	C7H14O	0.811														4
1-Heptene:	C7H14	0.6927	-0.88	0.34	-3.4	2.194	3.94	125	-340	(6.2)	20	1.582	3.8	(8.75)	110	1,13,4
cis-2-Heptene:	C7H14	0.703														4
trans-2-Heptene:	C7H14	0.696														22
cis-3-Heptene:	C7H14	0.697														4
trans-3-Heptene:	C7H14	0.693														22
n-Heptylbenzene:	C13H20	0.8567														20
Hexachlorobenzene:	C6Cl6															4,16
Hexachlorobutadiene:	C4Cl6	1.682														4
1,2,3,4,5,6-Hexachloro-cyclohexane:(a-isomer):	C6H6Cl6															20
(g-isomer): (Lindane)	C6H6Cl6	1.87														20
Hexachlorocyclopentadiene:	C5Cl6	1.701														4,15
Hexachloroethane:	C2Cl6	2.091														15
Hexacosanoic acid:	C26H52O2	0.820(100)														20
n-Hexadecane:	C16H34	0.775	-0.7	3.34	-32.6	2.144	3.6	150	-200	8(300)	13	2.77(300)	2.9	26(300)	100	13,4
Hexadecanoic acid:	C16H32O2	0.8534		7.1(75)		2.267		194	-511							4,21
1-Hexadecanol:	C16H34O	0.8116		36.8	n=-11.8	2.17	4.1	267	207	4.1	14			7.6	58	4,15
1,5-Hexadiene:	C6H10	0.6923		0.266	-1.31											15
2,4-Hexadien-1-ol:	C6H10O	0.871														19,20
Hexaldehyde: (Hexanal)	C6H12O	0.8139														19,20
Hexamethylbenzene:	C12H18	0.8(200)	-0.8	0.32(200)	-1.8	2.45(200)	3.6	110(200)	-140							13
Hexamethylene diamine:	C6H6N2															4
Hexamethyleneimine:	C6H13N	0.88														10,4
Hexamethylenetetramine:	C6H12N4	1.331														16
Hexamethyl phosphoramide:	C6H18ON3P	1.027		3.47												14,4
n-Hexane:	C6H14	0.659	-0.94	0.2926	-3.1	2.227	4.1	120	-340	6.57	21.8	1.71	4.15	10.7	110	13,4,16
		0.613*		0.202*		2.39*		100.4*		7.3*		1.91*		16.9*		
2,5-Hexanediol:	C6H14O2	0.961														20
2,5-Hexanedione:	C6H10O2	0.973														20
Hexanoic acid:	C6H12O2	0.929	-0.9	2.833	-73	2.15	1.32	145	-180	6.6	21	1.498	2.3	5.5	100	13
2-Hexanone:	C6H12O	0.8209		0.584												10,15,4
1-Hexene:	C6H12	0.668	-0.94	0.25	-2.75	2.18	3.8	121	-260	(7.1)	22	1.582	3.8	(10)	120	13,1,4,14
cis-2-Hexene:	C6H12	0.681														4,19
trans-2-Hexene:	C6H12	0.669														4
cis-3-Hexene:	C6H12	0.674														19
Trans-3-Hexene:	C6H12	0.671														19
Hexyl acetate:	C8H16O2	0.876														15
n-Hexyl alcohol: (1-hexanol)	C6H14O	0.816	-0.72	5	-86.3	2.23	11.8	143	-120	(6.03)	25	(1.63)	3.1	(6.75)	110	13,4
2-Hexyl alcohol: (2-hexanol)	C6H14O	0.81														15
3-Hexyl alcohol: (3-hexanol)	C6H14O	0.814														15
Hexylamine: (1-hexanamine)	C6H15N	0.7671														4
Hexylbenzene:	C12H18	0.8595	-0.31	1.52	-31	1.845	2.9	129	-160							13,4
Hexyl chloride:	C6H13Cl	0.879														20
Hexylcyclohexane:	C12H24	0.807	-0.7	2.21	-28.6	1.95	4	109	-120							13
Hexylcyclopentane:	C11H22	0.793	-0.7	1.41	-16.1	1.8	3.4	126	-190							13
Hexylene glycol:	C6H14O2	0.9234														10,4

continued

Table Db-1 Selected Physical and Termal Properties of Gases and Liquids for Sizing Equipment (at 25°C unless noted)

	Formula	LIQUID								GAS						
		dens Mg/m	dD/dT x1000	vis mPa.s	dv/dT x1000	cp kJ/kg.K	dc/dT x1000	k mW/m.K	dk/dT x1000	vis µPa.s	dv/dT x1000	cp kJ/kg.K	dc/dT x1000	k mW/mK	dk/dt x1000	References
Histidine:	C9H9N3O2															
Homopiperazine:	C5H12N2	1.0335														20
Hydrazine:	H4N2	1	-0.9	0.92	n=-4	3.09	2	314	-150	8.6	30	1.63	3	12.9	100	2,7,4
Hydrogen:	H2									8.92	19.9	14.3	3.6	181	400	13,4
		0.071*		0.0127*		9.74*		119*		1.12*		11.7*		16.3*		
/carbon dioxide mix: 40% H2														47		3,9
/nitrogen mix: 75% H2										15	24			100		3,9
/nitrous oxide mix: 40% H2														47		3,9
Hydrogen bromide:	HBr	2.13(-50)		0.5(-50)		0.749(-50)		131(-50)		18.5	64	0.36	0	9	30	13,4
-hydrobromic acid, 48%:		1.486														3
Hydrogen chloride:	HCl	1.24(-100)		0.58(-100)		1.60(-100)		421(-100)		14.5	52	0.795	0	14	50	13,4
		1.19*	-2.8*	0.407*	-4.26*	1.61*	6*	337*	-1267*	9*		0.85*		8.6*		
-hydrochloric acid, 30%wt:	HCl	1.48		1.192	-12.5	2.03	2.5	460								3
-hydrochloric acid, 10%wt:				1	-3.7	2.62	4.2									
Hydrogen cyanide:	HCN	0.681	-1.2	1.81	-22		1.7	289	-900	7.4	31.9	1.34	1.3	12.5	60	1,13,4
Hydrogen fluoride:	HF	0.941	-2.42	0.099	-6.28	3.9	12.6	427	-1100	12.5	41	1.457	0	26	93	13,4
		0.968*		0.215*		3.04*		402*		10.9*		1.46*		21*		
-hydrofluoric acid, 35%:		1.15														3
Hydrogen iodide:	HI	2.86(-50)		1.42(-50)		0.46(-50)		173(-50)		18.9	65	0.226	0.05	6	27	13,4
Hydrogen peroxide:	H2O2			1.155	-8.9	2.514	2.14	354	-470	22.8	45			19	80	7,4
Hydrogen sulfide:	H2S	0.98(-50)		0.24(-50)		1.875(-50)		211(-50)		12.7	42	1.005	0.38	18	60	13,4
		0.965*		0.423*		1.83*		233*		9.2*		1.02*		9.1*		
Hydroquinone:	C6H6O2															4
N-(2-Hydroxyethyl)morpholine:	C6H13NO2	1.083														20
N-(2-Hydroxyethyl)piperidine:	C7H15NO	0.973														20
Hydroxylamine: (Oxammonium)	H3NO															4
I*																
Indane:	C9H10	0.9639														15
Indene:	C9H8	0.9968														15,4
Iodine:	I	3.78(150)		1.8(150)		0.318(150)		112(150)		21.8(200)	46	0.147(200	0.01	5(200)	10	13,4
Iodobenzene:	C6H5I	1.825	-1.52	1.58	-26	0.777	0.42	100	-100							13,4
Iodoform:	CHI3	4.008														20
1-Iodopropane:	C3H7I	1.7489		0.71	-21											15,3
2-Iodopropane:	C3H7I	1.703		0.66	-21											15,3
Isoamyl acetate:	C7H14O2	0.8664	-0.89	0.7895		1.923		130								10
Isoamyl alcohol:	C5H12O	0.8053	-0.94	3.902	-80	2.396	10	132	-170							13,4
sec-Isoamyl alcohol:	C5H12O	0.8138		3.51												14
Isobutane:	C4H10	0.551	-2	0.145	-1.45	2.45	7.7	87	-230	7.58	25	1.67	4.8	16	110	13,4
Isobutanol:	C4H10O	0.8103	-0.85	3.527	-74.8	2.373	12.4	108	-180	7.3	24	1.513	3.28	9	120	13,14,4
Isobutyl acetate:	C6H12O2	0.8695	-1.05	0.651												10,14,4
Isobutyl acrylate:	C7H12O2	0.884														10,4
Isobutylamine:	C4H11N	0.7297		0.553		3.09										10,14,4
Isobutyl benzene:	C10H14	0.853														4,22
Isobutylene:	C4H8	0.5898	-1.1			2.3	3.05			8.01	27.4	1.59	3.46			1,4
Isobutylene oxide:	C4H8O	0.801														22,15
Isobutyl isobutyrate:	C8H16O2	0.8542														14,4
Isobutyraldehyde:	C4H8O	0.7836														10,4
Isobutyric acid:	C4H8O2	0.9429		1.23	-21											14,15,4
Isobutyronitrile:	C4H7N	0.7656		0.487	-6.3											14,4
Isodecyl alcohol:	C10H22O	0.835														4,22
Isooctyl alcohol:	C8H18O	0.827														10
Isopentane:	C5H12	0.615	-0.95	0.2018	n=-2.8	2.285	4.1	108	-386	7.25	22	1.708	4.3	13	120	13,4
Isophorone:	C9H14O	0.921		2.52	-42.3	1.9		151				1.297				10,4
Isoprene:	C5H8	0.6759	-0.9	0.2	n=-3	1.758	2.4	119	-140	7	28	1.54	3.7	13	103	1,0 7,4
Isopropanolamine:	C3H9NO	0.973														15,4

Isopropyl acetate:	C5H10O2	0.86	-1.3	0.52	-4			136	-4180	8	28	1.17	3.3	10	84	2,4
Isopropyl alcohol:	C3H8O	0.7813	-0.74	2.078	-70.76	2.565	13.8	141	-90	7.193	31	1.556	3.4	13	120	13,4
Isopropylamine:	C3H9N	0.7135	-1.08	0.314	-7.11	2.78	1.77	142	-155	7.7	24	1.69	4	10.5	140	13,14,4
Isopropyl benzoate:	C10H12O2	1.005														22,23
Isopropyl chloride:	C3H7Cl	0.856		0.314	-1.6											4,22
Isopropylcyclohexane:	C9H18	0.799	-0.78	0.9566	-9.9	1.88	4.2	104	-100							1, 13
Isopropylcyclopentane:	C8H16	0.771														22,19
Isopropyl ether:	C6H14O	0.7182		0.379		2.126										14, 10
Isovaleraldehyde:	C5H10O	0.803														15
Isovaleric acid:	C5H10O2	0.937		2.41												4
K*																
Kerosene: (No. 1 Fuel Oil)				2.3	-18.5			149	-155							
Ketene:	C2H2O	0.826	-2	1.1(-50)		2.32	4	182	-680	11.5		1.143		17		13
Krypton:	Kr									25.2	73	0.247	0	10	30	13
L*																
Lactic acid:	C3H6O3	1.206		40.33												15
Lactose:	C12H22O11															20
Linoleic acid:	C18H32O2	0.902														20
Lysine:	C6H14N2O2															4,15
M*																
Maleic acid:	C4H4O4	1.59		8.7	-76	1.657	3	192	-253	6.61	24			28		4
Maleic anhydride:	C4H2O3	1.48		1.5(70)		1.67										4
Malic acid:	C4H6O5	1.21														T
Mercuric chloride:	HgCl2	5.44														
Mercuric oxide:	HgO	11.14														
Mercury:	Hg	13.534	-2.44	1.52	-5.7	0.139	-0.04	8513	3340	6.62(400)	20	0.104(400	0	11(400)	40	13,1
		12.73*		0.884*		0.136*		122*		61.7*		0.104*		10.4*		
Mesitylene:	C9H12	0.861	-0.8	0.688	-9.3	1.744	3	135	-240	(6.4)	18	(1.34)	2.97	(7.25)	90	13,4
Mesityl oxide:	C6H10O															4
Metaldehyde:	(C2H4O)4															19
Methacrylaldehyde:	C4H6O	0.841														4
Methacrylic acid :	C4H6O2	1.009		1.32(20)												4
Methane: (Marsh gas).	CH4									11.2	30	2.24	3	33.7	190	13,1,4
		0.409*	-1.7*	0.106*	-0.9*	2.23*	2.84*	193*	-1500*							
Methanol: (Methyl alcohol)	CH4O	0.7875	-0.9	0.5399	n=-4.2	2.524	5.9	200	-300	9.65	34	1.424	2.28	11.75	190	13,1,4
		0.751*		0.326*		2.88*		191.4*		11.1*		1.55*		18.3*		
- 90%:				0.76	n=-4.1											3
-40%				1.86	n=-7.9											3
Methionine:	C5H11NO2S															
1-Methoxy-1,3-butadiene:	C5H8O	0.826														22,23
3-Methoxybutanol:	C5H12O2	0.9														22,23
3-Methoxybutyl acetate:	C7H14O3	0.949														22,23
3-Methoxybutyraldehyde:	C5H10O2	0.921														22,23
3-Methoxypropionitrile:	C4H7NO	0.934														4
Methyl acetate:	C3H5O2	0.9275	-1.3	0.3699	-3.82	2.14	4.54	155	-320	7.575	27	1.316	2.2	12.3	90	13,4
		0.875*		0.226*		1.92*		157*		8.9*		1.19*		14.2*		
Methyl acetoactate:	C5H8O3	1.0724	-0.76	1.704												14
Methyl acetylene:	C3H4	0.626	-0.94	0.07	-2.4	2.6	3.6	115	-600	8.86	28	1.516	3.02	19	100	1,13
Methyl acrylate:	C4H6O2	0.95		0.45	n=-3.6	1.76	0.44	176	-125	7.2	25	1.05	27	10.4	180	2,14,4
Methylal:	C3H8O2	0.854	-1.27	0.33												14,4
Methylamine:	CH5N	0.655	-1.14	0.14	-3.2	3.357	1.5	199	-340	9.6	31	1.665	4.13	18	120	4
N-Methylaniline:	C7H9N	0.982	-0.8	2.114	-37	2.045	1.07	158	-200							13,4
2-Methylaziridine:	C3H7N	0.8		0.417		1.77		151				1.49				10,14,4
Methyl benzoate:	C8H8O2	1.083	-0.97	0.6395	-5.5	1.594	2.8	146	-200							13,14
a-Methylbenzylamine:	C8H11N	0.948														15
a-Methylbenzyl-																

continued

Table Db-1 Selected Physical and Termal Properties of Gases and Liquids for Sizing Equipment (at 25°C unless noted)

	Formula	LIQUID								GAS						
		dens Mg/m	dD/dT x1000	vis mPa.s	dv/dT x1000	cp kJ/kg.K	dc/dT x1000	k mW/m.K	dk/dT x1000	vis µPa.s	dv/dT x1000	cp kJ/kg.K	dc/dT x1000	k mW/mK	dk/dt x1000	References
Dimethylamine:	C10H15N	0.899														22,23
Methyl borate:	C3H9BO3	0.93														23
Methyl bromide:	CH3Br	1.665	-2.56	0.261	-4.66	0.8355	0.26	98	-220	13.4	46	0.452	0.72	80	50	13,4
3-Methyl-1,2-butadiene:	C5H8	0.645														15
2-Methyl-1-butene:	C5H10	0.657														4
2-Methyl-2-butene:	C5H10	0.621														4
3-Methyl-1-butene:	C5H10	0.627(20)		0.216(0)		2.181						1.53				17.4
N-Methylbutylamine:	C5H13N	0.728														22,15,23
Methyl tert-butyl ether:	C5H12O	0.736	-1	0.298	-2.96	2.123	3.62	130	-320							15,23
		0.706*		0.231*		2.3*		108*		8.5*	31*	1.78*	6.25*	15.4*	102*	13,4
Methyl sec-butyl ketone:	C6H12O															24
Methyl tert-butyl ketone:	C6H12O															24
Methyl butyrate:	C5H10O2	0.891	-1.13	0.554	-4.98	2.03	2.4	139	-240							13,4
Methyl carbonate:	C3H6O3	1.1														23
Methyl chloride:	CH3Cl	1.05(-50)		0.259	-1.18	1.5(-50)		199(-50)		11	36	0.808	1.4	10.6	80	1,13,4
Methyl chloroacetate:	C3H5ClO2	1.2														4
Methylcyclohexane:	C7H14	0.7645	-0.9	0.6857	-9.86	1.882	3.82	111	-180	(6.9)	20	(1.51)	4.14	(3.5)	140	1,13,14,4
1-Methylcyclohexanol:	C7H14O	0.9251	-1.1													14,4
cis-2-Methylcyclohexanol:	C7H14O	0.9318	-0.85	18.08												14,4
trans-2-Methylcyclohexanol:	C7H14O	0.92085		37.13												14,4
cis-3-Methylcyclohexanol:	C7H14O	0.9111	-0.92	19.7												14,4
trans-3-Methylcylcohexanol:	C7H14O	0.9176	-0.76	19.5	-780											14,4
Methyl cyclopentadiene:	C6H8	0.9														15,23
Methylcyclopentadiene, dimer:	C12H16	0.941														20,15
Methylcyclopentane:	C6H12	0.744	-0.94	0.4785	-5.7	1.886	4.62	121	-360	7.2	22	1.34	4.46	10.75	110	13,14,4
N-Methyldibutylamine:	C9H21N	0.745														20
N-Methyldiethanolamine:	C5H13NO2	1.036				2.64		200								4
Methylene diiodide:	CH2I2	3.3254														15
N-Methylethanolamine:	C3H9NO	0.936														23,22
Methylethyl ether:	C3H8O	0.692														4
Methyl ethyl ketone:	C4H8O	0.7997	-1.06	0.384	-4.6	2.235	3.12	144	-260	7.33	25	1.447	2.83	9.5	100	13,14,4
2-Methyl-5-ethylpyridine:	C8H11N	0.916														22,23
N-Methylformamide:	C2H5ON	0.9988	-0.867	1.65												14,4
Methyl formate:	C2H4O2	0.97	-1.1	0.318	-5.4	2.04	3	184	-444	9.5	31	1.276	1.8	12.5	100	13,4
2-Methylheptane:	C8H18	0.698														15,4
3-Methylheptane:	C8H18	0.7														15,4
4-Methylheptane:	C8H18	0.699														15,4
2-Methyl-1-heptene:	C8H16	0.715														22,20,19
2-Methyl-2-heptene:	C8H16	0.719														19
3-Methyl-1-heptene:	C8H16	0.706														19
3-Methyl-2-heptene:	C8H16	0.724														22
4-Methyl-1-heptene:	C8H16	0.712														22
5-Methyl-1-heptene:	C8H16	0.711														22
cis-5-Methyl-2-heptene:	C8H16	0.718														22
6-Methyl-1-heptene:	C8H16	0.707														22
cis-6-Methyl-2-heptene:	C8H16	0.713														22
2-Methylhexane:	C7H16	0.6744		0.366	-2.16	2.188						1.66				14,4
3-Methylhexane:	C7H16	0.68295		0.351	-4.3	2.188						1.66				14,4
5-Methyl-2-hexyl alcohol:	C7H16O	0.808														20
Methyl hexyl ketone:	C8H16O	0.814														20
Methyl Hydrazine:	CH6N2	0.866														20
Methyl 3-hydroxybutyrate:	C5H10O3	1.053														20
Methyl iodide:	CH3I	2.265		0.479	-2.52	0.593						0.313				14,18,4
Methyl isoamyl ketone:	C7H14O	0.807														4,20
Methyl isobutyl ketone:	C6H12O	0.7961	-1.2	0.542	-4.6	2.13	2.8	138	-260	6.1	22	1.56	2.8	9.5	100	2,14,4
Methyl isocyanate:	C2H3NO	0.951														4,20
Methyl isopropenyl ketone:	C5H7O	0.847														23
Methyl isopropyl ketone:	C5H10O	0.798														4
Methyl mercaptan:	CH4S	0.864	-1.24	0.19	-4	1.842	0	187	-800	9.6	32	1.054	1.7	12	80	13,4

Methyl methacrylate:	C5H8O2	0.934	-1.16	0.55	-4.7	1.84	0.44	157	-125	7.9	25	1.088	2.7	9	180	2,14,4
N-Methylmorpholine:	C5H11NO	0.921		0.69												11
1-Methylnaphthalene:	C11H10	1.001														15,4
2-Methylnaphthalene:	C11H10	1														15,4
2-Methylnonane:	C10H22	0.7281														19
3-Methylnonane:	C10H22	0.7354														19
4-Methylnonane:	C10H22	0.7323														19
5-Methylnonane:	C10H22	0.7326														19
2-Methyloctane:	C9H20	0.708														19,4
3-Methyloctane:	C9H20	0.716														19,4
4-Methyloctane:	C9H20	0.715														4
2-Methyl oleate:	C19H3602	0.8702	-0.72	5.15	-54											14,4
2-Methylpentanal:	C6H12O	0.805														19
2-Methylpentane:	C6H14	0.6485	-0.94	2.855	-2.9	2.262	5.02	113	-380	6.8	20	1.698	4.27	11	120	13,14
3-Methylpentane:	C6H14	0.66	-0.92	0.31	-2.8	2.211	4.18	113	-340	6.8	2	1.739	4.2	11	120	13,14,4
2-Methyl-1,3-pentanediol:	C6H14O2	0.969														23
2-Methyl-1,5-pentanediol:	C6H14O2	0.97														
2-Methylpentanoic acid:	C6H12O2	0.918														15
2-Methylpentanol:	C6H14O	0.8206														14,4
2-Methyl-2-pentanol:	C6H14O	0.811														19
2-Methyl-3-pentanol:	C6H14O	0.82														19
3-Methylpentanol:	C6H14O	0.819														19
3-Methyl-2-pentanol:	C6H14O	0.825														19
3-Methyl-3-pentanol:	C6H14O	0.824														15
4-Methylpentanol:	C6H14O	0.809														19
4-Methyl-2-pentanol:	C6H14O	0.808														19,4,15
2-Methyl-2-pentenal:	C6H10O	0.86														20,19
2-Methyl-1-pentene:	C6H12	0.6799														4
2-Methyl-2-pentene:	C6H12	0.6865														19
3-Methyl-1-pentene:	C6H12	0.67														19
4-Methyl-1-pentene:	C6H12	0.665														19
cis-4-Methyl-2-pentene:	C6H12	0.671														19
trans-4-Methyl-2-pentene:	C6H12															4
N-Methylpiperazine:	C5H12N2	0.898														20
2-Methyl-2-propanethiol:	C4H10S	0.7947		0.46531		1.94										8
Methyl propionate:	C4H8O2	0.909	-1.2	0.55	-7.1	1.97	2.4	145	-280							13,4
Methyl propyl ether:	C4H10O	0.719	-1.1	0.244	-3.1	2.231	2.9	133	-378							13,15
Methyl propyl ketone:	C5H10O	0.801		0.48	-21											4
N-Methylpyrrolidone:	C5H9NO	1.02		1.65		1.68		95								14,4
N-Methyl-2-pyrrolidone:	C5H9NO	1.0279		1.666												14
Methyl salicylate:	C8H8O3	1.177	-0.57	2.01	-26	1.59	2.2	148	-190							13,14
a-Methylstyrene:	C9H10	0.9082														19,4
2-Methyltetrahydrofuran:	C5H10O	1.191		0.438	-6.5											15,23
4-Methyltetrahydropyran:	C6H11O	0.953														20
2-Methylthiophene:	C5H6S	1.012														20
3-Methylthiophene:	C5H6S	1.015														20,15
Methyl valerate:	C6H12O2	0.875		0.713												15
Methyl vinyl ketone:	C4H6O	0.842														20
Monoethanolamine:	C2H7NO	1.012	-0.79	19.346	-1093	2.086										14
Morpholine:	C4H9NO	0.99547	-0.948	2.039	-49	1.996										14,10,4
N*																
Naphtha: petroleum		0.7348		4.477		2.05		150	0							10
Naphthalene:	C10H8	1.024	-0.82	1.155	-5.1	1.523	3.75	148	-140	11.6(300)	19	1.89(300)	2.11	28(300)	90	13,4
1-Naphthol:	C10H8O															20
2-Naphthol:	C10H8O															20
Neon:	Ne									31.2	66	1.03	0	49	110	13
		1.205*		0.127*		1.87*		113*		4.63*		1.31*		7.7*		
Nicotinic acid:	C6H5NO2															20
Nitric acid:	HNO3	1.502		0.809	-4.7	1.506	-1.93	281	540							4
90%wt:		1.4826		2.28	n=-5.1	2.16		310								3,11,26

continued

Table Db-1 Selected Physical and Termal Properties of Gases and Liquids for Sizing Equipment (at 25°C unless noted)

	Formula	LIQUID								GAS						References
		dens Mg/m	dD/dT x1000	vis mPa.s	dv/dT x1000	cp kJ/kg.K	dc/dT x1000	k mW/m.K	dk/dT x1000	vis µPa.s	dv/dT x1000	cp kJ/kg.K	dc/dT x1000	k mW/mK	dk/dt x1000	
60%wt:		1.3667		1.12	n=-3.7	2.68		400								3,11,26
Nitric oxide:	NO									19.2	48	0.971	0.09	26	70	13,7
p-Nitroaniline:	C6H6NO2	1.437														23
Nitrobenzene:	C6H5NO2	1.198	-0.98	1.8	-24	1.504	2.9	150	-14							1,13,14,4
2-Nitrobiphenyl:	C12H9NO2	1.2														20,23
4-Nitrobiphenyl:	C12H9NO2															16,20
o-Nitrochlorobenzene:	C6H4ClNO2	1.348														20
p-Nitrochlorobenzene:	C6H4ClNO2	1.298														20
Nitroethane:	C2H5NO2	1.0446	-1.4	0.638	n=-3	1.847	1.5	167	-320	7.5	24	1.088	1.8	12.5	110	2,14,4
Nitrogen:	N2								-1640	17.8	49	1.038	0.02	26	72	1,13,4
		0.807*		0.163*		2.064*		137*		5.41*		1.123*		7.54*		
Nitrogen dioxide:	NO2	1.434	-2.4	0.403	-3.65	1.5425	1.5	128	-500	14.9	42	0.808	0.66	11.8	-320	13,7
Nitrogen peroxide:	N2O4									17	42	0.858		11.8	-320	13
Nitroglycerine:	C3H5N3O9															4
Nitromethane:	CH3NO2	1.129	-1.38	0.6205	-7.3	1.737	1.5	202	-320	7.6	24	0.94	1.63	12.5	110	13,2,14,4
2-Nitrophenol:	C6H5NO3	1.495														15
4-Nitrophenol:	C6H5NO3	1.495														15
1-Nitropropane:	C3H7NO2	0.9961	-1.07	0.79	-10.4	1.972										14,4
2-Nitropropane:	C3H7NO2	0.9829	-1.099	0.721	-9.3											14,4
4-Nitrosodiphenylamine:	C12H10N2O															20
N-Nitrosodipropyl amine:	C6H14N2O	0.916														19
o-Nitrotoluene:	C7H7NO2	1.1582	-0.96	2.129	-48.2	1.157	3.6	139	-400							13,4
m-Nitrotoluene:	C7H7NO2	1.1525	-0.9	2.23	-19.7	1.119	4.3	140	-163							13,4
p-Nitrotoluene:	C7H7NO2	1.151	-0.92	1.177	-5.56	1.213	1.92	151	-180							13,4
Nitrous oxide:	N2O									14.9	47	0.879	0.92	17.5	90	13,7,4
n-Nonadecane:	C19H40	0.782	-0.67	2.86(50)		2.28(50)		153	-185	8.2(400)	23	3.05(400)	4.57	31(400)	180	1,13,4
Nonadecanoic acid:	C19H38O2	0.8771														4
n-Nonane:	C9H20	0.713	-0.78	0.666	-9.44	2.193	3.26	131	-240	(5.6)	16	(1.819)	3.5	(6.75)	110	13,4
		0.614*		0.213*		2.72*		95*		7.4*		2.24*		20.8*		
Nonanoic acid:	C9H18O2	0.906														4,15
2-Nonanone:	C9H18O	0.8208														19
3-Nonanone:	C9H18O	0.821														19
4-Nonanone:	C9H18O	0.819														20
5-Nonanone:	C9H18O	0.826														19
1-Nonene:	C9H18	0.72531	-0.788	0.586	-72	2.156						1.596				10,14,4
n-Nonyl alcohol: (1-Nonanol)	C9H20O	0.823														4,15
2-Nonyl alcohol: (2-Nonanol)	C9H20O	0.819														
n-Nonyl benzene:	C15H24	0.851														4
1-(n-Nonyl) naphthalene:	C19H26	0.932														4
Nonylphenol:	C15H24O	0.9494		1513		2.034		150	0							10,4
O*																
Octacosanoic acid:	C28H56O2					2.21										5,25
n-Octadecane:	C18H38	0.78	-0.7	3.16	-27	2.2	3.04	146	-180	8.6(400)	24	3.05(400)	4.6	33(400)	190	1,13,14,15,4
Octadecanoic acid:	C18H36O2	0.847(70)		9.05(75)				194	-853							23,4
1-Octadecanol:	C18H38O	0.812(58)														4
1-Octadecene:	C18H36	0.791														4
9-Octadecenoic acid:	C18H34O2	0.8895	-0.68	27.64	-465	2.2		224	-495							10,14,4
9-Octadecen-1-ol:	C18H360	0.849														21,15
Octadecylamine:	C18H39N	0.777														15
n-Octane:	C8H18	0.6985	-0.84	0.5151	-4.7	2.2	4.05	129	-300	8.45	19.5	2.437	3.5	26	120	1,13,4
		0.611*		0.203*		2.5*		98*		7.4*		2.11*		18.4*		
n-Octanoic acid:	C8H16O2			4.69												4,15
1-Octanol:	C8H18O	0.8257	-0.66	7.27	-332	2.26	7.88	136	-140	(6.1)	23	(1.659)	3.19	(6)	120	13,14,4
2-Octanol:	C8H18O															4
1-Octene:	C8H16	0.711	-0.86	0.44	-5.3	2.146	3.46	127	-150	(5.175)	19	(1.78)	3.26	(8.75)	110	1,13,4

cis-2-Octene:	C8H16	0.719														19
trans-2-Octene:	C8H16	0.714														4,19
cis-3-Octene:	C8H16	0.716														19
trans-3-Octene:	C8H16	0.71														4,19
cis-4-Octene:	C8H16	0.716														19
trans-4-Octene:	C8H16	0.715														4,19
n-Octylbenzene:	C14H22	0.851														22
tert-Octyl mercaptan:	C8H18S	0.843														4,20
Oils, crude:		0.7		4.477	-474	1.984		131	0							10
Oils,edible: Canola:		0.917		69	-644	2.043		171								10
Oils, edible: Castor:		0.9597		708	-7026	2		180								23,10
Oils, edible: Coconut:		0.9198		36.7	-411	2.01	0	166	0							10
Oils, edible: Corn:				50												
Oils, edible: Cottonseed:		0.913		1142	-80850	2.16		131	0							10
Oils, edible: Fish		0.913		1142	-80850	2.001		131	0							10
Oils, edible: Olive:		0.911		68	-657	2.001		168	-13							23,10
Oils, edible: Palm:		0.9086				2.01		131	0							10
Oils, edible: Peanut:		0.8974		1142	-80850	2.097		170	0							10
Oils, edible: Safflower:																10
Oils, edible: Soya bean:		0.8942		52	-485	1.921		170	0							10
Oils, edible: vegetable:		0.923		1140	-80850	2.001		170	0							10
Oils, linseed:		0.929		38	-343	1.93		167	0							10
Oils, lubricating:																10
Oils, mineral:																10
Oils, sperm:		0.879		30	n=-8.6	1.98		131	0							10
Oxalic acid:	C2H2O4	1.9														15
Oxalic acid, dihydrate:	C2H2O4.2H2O	1.653														15
1,4-Oxathiane:	C4H8OS	1.114														20
Oxygen:	O2									20.3	51	0.913	0.25	26	80	13,4
		1.135*		0.196*		1.63*		148*		6.85*		0.96*		8.5*		
P*																
Paraformaldehyde:	(CH20)x	0.998														
Paraldehyde:	C6H12O3	0.988	-1.18	1.128	-10.1	1.951	3.6	144	-125	(7.5)	24	(1.409)	2.6	(10.5)	100	13,4
Pentachloroethane:	C5HCl5	1.67	-1.58	2.37	-42	0.86	0.4	99	-180							13,4
Pentachlorophenol:	C6HCl5O	1.978														
n-Pentadecane:	C12H32	0.764	-0.7	2.733	-26	2.168	3.48	140	-200	8.3(200)	29	2.77(200)	5.6	28(300)	100	13,4
Pentadecanoic acid:	C15H30O2															4
Pentadecanol:	C15H32O	0.817		2.19	-28	2.1	0	150	0			1.578				10
1,2-Pentadiene:	C5H8	0.609	-2.49	0.2	-3.1	2.216	3.13	121	-400							13,4
trans-1,3-Pentadiene:	C5H8	0.671	-0.93	0.2	-3.1	2.187	3.6	121	-400	7.3	24	1.553	3.52	13.5	100	13,4
1,4-Pentadiene:	C5H8	0.655	-1.07	0.2	-3.1	2.184	3.82	121	-400							13,4
2,3-Pentadiene:	C5H8	0.69	-0.91	0.2	-3.1	2.234	2.69	109	-644							13
Pentamethylbenzene:	C11H16	0.916	-0.76	1.05	-5.2	1.723	4.1	136	-180							13
Pentane:	C5H12	0.6214	-0.98	0.225	-3	2.29	3.3	135	-178	6.875	23	1.71	4.19	11.25	130	1,13,14,4
		0.61*		0.196*		2.34*		107*		6.9*		1.79*		16.7*		
1,5-Pentandiol:	C5H12O2															4
Pentanoic acid:	C5H10O2	0.9378	-0.84	2	-44.8	2.06	4.4	145	-180							13,14,4
Pentatriacontanoic acid:	C35H70O2															5
1-Pentene:	C5H10	0.6353	-1.034	0.198	-2.2	2.216	4.29	117	-422	6.6	26.2	1.66	3.73	12	120	13,4
cis-2-Pentene:	C5H10	0.6502	-1.06			2.165	3.48									1,14
trans-2-Pentene:	C5H10	0.6431	-1.024			2.244	3.76									1,14
3-Pentyl alcohol:	C5H12O	0.816		3.35	-268	2.85										14,4
Pentylbenzene:	C11H16	0.855	-0.72	1.21	-23	1.834	2.8	129	-180							13,4
n-Pentylcyclohexane:	C11H22	0.801	-0.72	1.56	-32.3	1.94	4	108	-100							13
Pentylcyclopentane:	C10H20	0.7873	-0.74	1.067	-17.56	1.777	3.4	126	-200	(5.7)	17	(1.638)	2.5	(6.25)	90	13
Pentyl mercaptan:	C5H12S	0.8376	-31	0.504		1.133										17
1-Pentyne:	C5H8	0.6901														
Peracetic acid:	C2H4O3	1.15		3.28	-94											10,4

continued

Table Db: Liquid and Gases Properties of Chemicals at 25 oC (unless noted)

	Formula	LIQUID								GAS						
		dens Mg/m	dD/dT x1000	vis mPa.s	dv/dT x1000	cp kJ/kg.K	dc/dT x1000	k mW/m.K	dk/dT x1000	vis µPa.s	dv/dT x1000	cp kJ/kg.K	dc/dT x1000	k mW/mK	dk/dt x1000	References
Perfluorodimethyl-cyclohexane:	C8F16	1.828														19,20
Perfluoromethyl-cyclohexane:	C7F14	1.7878														19
Perfluoro-n-heptane:	C7F16	1.733														19
Phenanthrene:	C14H10	1.179														
o-Phenetidine:(2-)	C8H11NO	1.556		0.51	n=-13.2											19,3
m-Phenetidine:(3-)	C8H11NO			14												19
p-Phenetidine:(4-)	C8H11NO	1.528		10.3	n=-13.1											19,4
Phenetole:	C8H10O	0.9605	-0.934	1.14	-21	1.869289										14,4
Phenol:	C6H6O	1.08	-1.19	4.1	n=-10.5	2.2	1.76	163	-260	(7.15)	26	(1.22)	2.42	(7.5)	100	13,14,4
		0.955*		0.351*		2.55*		175*		12.8		1.63*		28.9*		
Phenyl acetate:	C8H8O2	1.07														15
N-Phenylethanolamine:	C8H11NO	1.085														20
Phenyl ether:	C12H10O	1.071	-0.988	2.71	-27	1.58										1,14
Phenylhydrazine:	C6H8N2	1.099														20,4
N-Phenylpiperazine:	C10H14N2	1.062														20
Phosgene:	COCl2	1.378	-2.31	0.61	-21	1.02	0	132	-240	9.2	33	0.615	0.56	10	66	13,4
Phosphoric acid, liquid-																
-105%wt H3PO4:		1.93		780												3,11,26
-115%wt H3PO4:		2.06		2060		1.5										3,11,26
-85%wt H3PO4:		1.69		48	-550	1.75										3,11,26
-75%wt H3PO4:		1.57		19	-195	2.1										3,11,26
Phthalic acid:	C8H16O4															4
Phthalic anhydride:	C8H14O3															4
Phthalide:	C8H6O2	1.164														
a-Picoline:	C6H7N	0.95(15)				1.817	0									10,4
b-Picoline:	C6H7N	0.961(15)														4
q-Picoline:	C6H7N	0.957(15)														4
Picric acid:	C6H3N3O7															20
Pinacolyl alcohol:	C6H14O	0.812														20
Pinene:	C10H16	0.8539		1.4												14,4
Piperidine:	C5H11N	0.856	-0.92	1.17	-63	2.147	2.24	124	-260							13,14,4
Pivaldehyde:	C5H7O	0.793														20
Polyglycolamine, H221M:		1.004													6.25	
Potassium:	K	0.647(825	-0.23	0.135(825	0.16	0.819(825	0.05	27800(82	-29000	18.4(825)	21	0.934(825	-1.1	21.8(825)	6.25	1
	K	0.659*				0.805*				17.8*		1.113*		24.8*		1
Potassium benzoate:	C7H5KO2															19
Potassium chloride:	KCl															19
solution:						3.22										3
Potassium hydroxide:	KOH															19
solution:						3.14										3
Potassium nitrate:	KNO3															19
Potassium sulfate:	K2SO4															19
Potassium xanthate:	C3H5KOS2															19
Propadiene: (Allene).	C3H4	0.682(-50	-2.18	0.23(-50)	-4.6	2.02(-50)	3	170(-50)	-720	8.5	27	1.474	3.3	16	110	1,13,4
Propane:	C3H8	0.59(-50)	-1.1	0.20(-50)	-4.7	2.2(-50)	2.9	139(-50)		8.15	26	1.67	4.7	18	120	13,4
		0.582*		0.208*		2.24*		134*		6*		1.37*		10.7*		
1,3-Propanediol:	C3H8O2	1.05	-0.63	44.2	-1680	2.498	5.9	201	0							13,14
2-Propanethiol:	C3H8S	0.82														20
Propanoic acid:	C3H6O2	0.9878	-1.04	1.64	n=-4.1	2.187	4.39	149	-17	(7.65)	26	(2.541)	4.02	(18.5)	100	13,14,4
1-Propanol:	C3H8O	0.8	-0.78	1.93	-59.5	2.38	7	156	-140	6.98	30.9	1.529	3.68	8	120	1,13,14,4
b-Propiolactone:	C3H4O2	1.145		0.768		1.775		151			1.185					10,4
Propionaldehyde:	C3H6O	0.7912	-1.17	0.316	-4.05	2.184		159	-385			1.35				1,10,14,4
Propionic anhydride:	C6H10O3	1.0057	-1.13	1.061	-11.7	1.78	2.1	131	-60	(7.1)	24	(1.27)	2.3	(5.75)	110	13,14,4
Propionitrile:	C3H5N	0.777	-1.04	0.4106	-4.28	2.176	2.68	171	-260	6.98	23	1.33	2.68	11	80	13,14,4
Propyl acetate:	C5H10O2	0.883	-1.1	0.551	n=-4.1	1.922	2.4	137	-220	7.48	23	(1.38)	2.56	(6.75)	110	13,14,4
n-Propylamine:	C3H9N	0.7121	-1.04	0.353		2.73						1.628				14,10,4

n-Propylbenzene:	C9H12	0.8554	-0.92	0.791	-11.98	1.783	2.86	127	-200	(6.3)	19	(1.41)	2.97	(5.5)	100	13,4
n-Propyl chloride:	C3H7Cl	0.885	-1.21	0.2996	-5.87	1.657	1.43	116	-257	7.7	28	1.119	2.37	14	50	13,14,4
n-Propylcyclohexane:	C9H18	0.788														4
n-Propylcyclopentane:	C8H16	0.773	-0.8	0.642	-8.2	1.932	3.4	126	-220	(6.35)	18	(1.54)	3.77	(4)	120	13,4
Propylene: (Propene)	C3H6	0.5505	-0.1	0.19(-50)	-3.56	2.315	1.84	120	-340	8.6	29	1.52	3.76	17	120	1,13,4
		0.611*		0.151*		2.39*		119*		6.62*		1.31*		9.52*		
Propylene-1,2-carbonate:	C4H5O3	1.189		3.0		1.42		120								15
sec-Propylene chlorohydrin:	C3H7ClO	1.111														20
Propylenediamine:	C3H10N2	0.87														20
Propylene dichloride:	C3H6Cl2	1.156		0.7												20,15
1,2-Propylene glycol:	C3H8O2	1.0328	-0.72	46.5	-1900	2.48						1.611				14,4,T
Propylene glycol	C3H8O2															11
acrylate:(mono)	C6H10O3	1														23
butyl ether: (mono)	C9H2OO2	0.8843		3.25												11
dimethyl ether:	C5H12O2	0.855		5.8		2.0		110								20
ethyl ether: (mono)	C5H12O2	0.8979		2.05												11
methyl ether: (mono)	C4H10O2	0.9234		1.75								1.5				10,T,11
methyl ether acetate:	C6H12O3	0.968														20
phenyl ether:(mono)	C9H12O2	1.063														15
isopropyl ether:(mono)	C6H14O2	0.879														15
propyl ether: (mono)	C6H14O2	0.8865		2.65												11
Proyleneimine:	C3H7N	0.8		0.417		1.77		151				1.49				10,14,4
Propylene oxide:	C3H6O	0.823	-1.2	0.28	-10.1	2.06	2.3	150	-422	8.9	29	1.34	2.33	12	120	13,14,4
Propyl formate:	C4H8O2	0.9	-1.16	0.485	-5.94	1.99	1.8	146	-280							13,14,4
4-Propylheptane:	C10H22	0.731														22
2-Propylheptyl alcohol:	C10H22O	0.828														22
Propyl mercaptan:	C3H8S	0.836		0.376	-4.45	1.93		151				1.218				10,20,4
Propyl propionate:	C6H12O2	0.877	-1.08	0.5	-6.3	2.01	2.24	135	-180							13,4
m-Propyltoluene:	C10H14	0.8744														19
o-Propyltoluene:	C10H14	0.861														19
p-Propyltoluene:	C10H14	0.8584														19
Propyl valerate:	C8H16O2	0.874		1.02												3
Pseudocumene:	C9H12	0.872	-0.8	0.956	-10.8	1.771	2.66	131	-200	(6.18)	19	(1.37)	2.9	(7.3)	90	13,4
Pyrene:	C16H10	1.271														15
Pyridine:	C5H5N	0.978	-1	0.886	-14.4	1.71	2.44	160	-320	(8.7)	20	(1.11)	2.6	(6.5)	100	13,14,4
2-Pyrrolidone:	C4H7ON	1.107	-0.81	13.3												14,4
Q*																
Quinoline:	C9H7N	1.089	-0.794	3.37	-270	1.55		146								14,10,4
R*																
S*																
Salicylaldehyde:	C7H6O2	1.1525	-0.979	3.05	-30	1.626	3.5	153	-188							13,14,4
Salicylic acid:	C7H6O3	1.443		2.41	n=-5.7											4,15
Silane:	SiH4	0.68								112		1.33952				15,4
Silicon:	Si	2.533MP		0.88MP		1.01MP		428MP								15
Silicon dioxide:	SiO2															
Silicon tetrachloride:	SiCl4	1.5	-2.5	0.44	-3.4	0.92092				90		0.535808		6.27		15
Sodium:	Na	0.755(825	-0.24	0.165(825	-0.15	1.27(825)	0.2	52100(82	-42670	25.64(825	14.5	2.53(825)	-2.88	49.3(825)	-12.5	1
				1.1	n=-1.8					27.5*		2.62*		49.2*		
Sodium acetate:	C2H3NaO2	1.528														19
Sodium benzoate:	C7H5NaO2															19
Sodium bicarbonate:	CHNaO3	2.159														19
Sodium bromide:	NaBr	3.203		2556	n=-6.											19
Sodium chlorate:	NaClO3	2.49														19
Sodium chloride:	NaCl	2.165		347	n=-4.2											19
-brine: 25%				2.15	n=-5.6			571								3
Sodium dodecyl sulfate:	C12H25NaO4S															20

continued

Table Db: Liquid and Gases Properties of Chemicals at 25 oC (unless noted)

	Formula	LIQUID dens Mg/m	dD/dT x1000	vis mPa.s	dv/dT x1000	cp kJ/kg.K	dc/dT x1000	k mW/m.K	dk/dT x1000	GAS vis µPa.s	dv/dT x1000	cp kJ/kg.K	dc/dT x1000	k mW/mK	dk/dt x1000	References
Sodium fluoride:	NaF	2.558														19
Sodium hydrogen phosphate:	Na2HO4P															
ortho .2H2O:		2.066														19
ortho .7H2O:		1.679														19
ortho .12H2O:		1.52														19
Sodium hydrosulfide:	NaHS															19
Sodium hydroxide:	NaOH	2.13														19
50%:		1.53		85	n=-14.6											3
Sodium nitrate:	NaNO3	2.261		17.5	n=2.7											19
Sodium nonyl sulfate:	C9H19NaO4S															
Sodium sulfate:	Na2O4S															19
Styrene:	C8H8	0.90122	-0.82	0.696	-11	1.733	3.52	137	-220	(7.7)	20	(1.31)	2.64	(5.5)	100	13,14,4
Styrene oxide:	C8H8O	1.0523														15
Succinic anhydride:	C4H4O3															4
Succinonitrile:	C4H4N2	1.55	-16	2.59(60)	-583	1.767										14
Sucrose:	C12H22O11	1.587														
Sulfanilamide:	C6H8N2O2S															15
Sulfathiazole:	C9H9N3O2S2															20
Sulfobromophthalein:																
Sulfolane:	C4H8O2S	1.266	-0.835	11.28	-199	1.5										14,10,4
Sulfur:	S2	(1.87)	0.72	(22.6)	-125	(.57)	0	(94)	340	17.6(500)	21.6	0.574(500	0.034	14(500)	18	13,4
Sulfur dioxide:	SO2	1.557(-50)		0.68(-50)	n=-2.6	1.38(-50)		242(-50)		12.8	46	0.607	0.76	10	50	1,13,4
Sulfur hexafluoride:	SF6									15.6	46	0.641	1.56	14	80	13
Sulfuric acid:	H2SO4	1.8491		25		1.415										T
110%:				42.5	n=8.6											3
98%:		1.84		22.5	n=8.8	1.4651		3635								3
60%:		1.5		6.08	n=5.6			4327								3
Sulfur trioxide:	SO3			1.73		3.14		247	-1230	13.5	43			11	46	4
T*																
Tallow:		0.937		52		2		150								10,15
a-Terpineol:	C10H17O															19
1,1,2,2-Tetrabromoethane:	C2H2Br			7.49	-460											15,4
Tetrachlorodifluoroethane:	C2F2Cl4	1.644	-1.95	1.21	-12.1	0.61		78(50)								13,14,4
Tetrachlorodiphenyl:	C12H6Cl4	1.445	-1	120	-1544	1.175	7	1025	-5000							1
1,1,2,2-Tetrachloroethane:	C2H2Cl2	1.5866	-1.66	1.62	-30.5	0.955	0.58	113	-200			0.593				13,14,4
Tetrachloroethylene:	C2Cl4	1.6128	-1.64	0.8553	-9.74	0.8907	2.34	109	-200	(12.8)	32	(0.643)	2.5	(6)	40	13,14,4
Tetracosanoic acid:	C24H48O2	0.82(100)														
Tetradecane:	C14H30	0.7593	-0.725	2.218	-20.2	2.2	3	144	-188	8.6(300)	14	2.77(300)	2.8	29(300)	100	13,4
Tetradecanoic acid:	C14H28O2	0.858(60)		5.83(70)												4
1-Tetradecene:	C14H28	0.767		1.78	-53	1.896	0	150	0			1.6				10,4
Tetradecylbenzene:	C20H34	0.851		6.89	-115	1.507		151								10,19
1,1,3,3-Tetraethoxypropane:	C11H24O4	1.12		0.92												23
Tetraethylene glycol:	C12H18O5	1.115		44.9												10,4
diacrylate:	C1422O7	1.11														20
dibutyl ether:	C16H34O5	0.9														23
diethyl ether:	C12H26O5	0.97														20
dimethyl ether:	C10H22O5	1.018														20,19
methyl ether:(mono)	C9H20O5	0.987														15
Tetraethylene pentamine:	C8H23N5	0.998														10,16
1,2,3,6-Tetrahydro-benzaldehyde:	C6H9CHO	0.94														15
Tetrahydrofuran:	C4H8O	0.8842	-1.01	0.46	-2.6	2.19	0.56	139	-114	8	25	1.55	3.45	12.6	84	2,14,10,4
1,2,3,4,Tetrahydro-naphthalene:	C10H12	0.9662		2.003	-5.19	1.648	0	131	0			1.15				14,10,4
Tetrahydropyran:	C5H10O	0.8772		0.764		1.82										14

Tetrahydropyran-2-methanol:	C6H12O2	1.0254		11(20)													23
Tetrahydothiophene:	C4H8S	0.99379	-0.9	0.971	-12.8												14,4
1,1,3,5-Tetramethoxyhexane:	C10H22O4																22
Tetramethyl ammonium chloride:	C4H12ClN																
1,2,3,4-Tetramethylbenzene:	C10H14	0.901	-0.78	1.2	-19.4	1.803	2.62	138	-200								13
1,2,3,5-Tetramethylbenzene:	C10H14	0.886	-0.76	0.838	-12.4	1.79	2.22	138	-200								13,4
1,2,4,5-Tetramethylbenzene:	C10H14	0.879	-0.8	0.535	-2.2	1.873	2.6	137	-180	9(200)	18	1.96(200)	2.8	23(200)		90	13
2,2,3,3-Tetramethylbutane:	C8H18																15
N,N,N,N-Tetramethyl-1,3-butadiamine:	C8H20N2	0.786															15,19
N,N,N,N-Tetramethyl-1,2-ethylenediamine:	C6H16N2	0.77															15
2,2,3,3-Tetramethylpentane:	C9H12O	0.7															15,23,4
N,N,N,N-Tetramethyl-1,3-propanediamine:	C7H18N2																15
Tetramethyl urea:	C5H12N2O	0.9687															15
2-Thiabutane:	C3H8S	0.8422															15
2-Thiapropane:	C2H6S																22
Thiodiglycol:	C3H10O2S	1.182															15
Thiophene:	C4H4S	1.058	-1.46	0.613	-8.02	1.46	1.2	136	-280			0.868					13,14,4
Thiophosgene:	CSCl2	1.506						127				0.578					10
Thiourea:	CH4N2S	1.17															15,16
Thymine:	C5H6N2O2																19
Toluene:	C7H8	0.8623	-0.92	0.551	-7.08	1.73	3.76	133	-160	6.98	21	(1.25)	2.89	(14.5)		100	1,13,4
		0.778*	-1.16*	0.251*	-1.34*	1.81*	3.8*	113*	-200*	9.0*	24*	1.13*	3.0*	11.1*	74*		
Toluene-2-4-Diisocyanate:	C9H6O2N2	1.21		4.59	-118	1.666		170	0								4,10
m-Toluidine:	C7H9N	0.988	-0.5	3.3	-112	2.028											14,4
o-Toluidine:	C7H9N	0.9943		3.39	-134	2.05		159									14,10,4
p-Toluidine:	C7H9N	0.98195	-0.803	2.63	-34												14
Triallylamine:	C3H15N	0.809															19
Triacontanoic acid:	C30H60O2																25
1,2,3-Triazole:	C2H3N3	1.8161															19
Tributylamine:	C12H27N																4
Tri-n-butyl phosphate:	C12H27O3P	0.972															15,23
1,2,4-Trichlorobenzene:	C6H3Cl3	1.446															15,4
1,1,1-Trichloroethane:	C2H3Cl3	1.33	-1.62	0.758	-16.4	1.081	0.94	101	-200	9.4	32	0.721	0.88	8.25		500	13,4
1,1,2-Trichloroethane:	C2H3Cl3																4
Trichloroethylene:	C2HCl3	1.449	-2.81	0.543	-11.4	0.9536	0.71	115	-243	10.9	37	0.648	0.63	7.5		60	13,4
Trichlorofluoromethane:	CCl3F	1.477	-2.2	0.395	-9.14	0.887	0.77	87	-285	11.3	31	0.569	0.67	8		40	1,13,4
2,4,6-Trichlorophenol:	C6H3Cl3O	1.49(75)															19
1,2,3-Trichloropropane:	C3H5Cl3	1.388															15,4
2,2,3-Trichloro-propionaldehyde:	C3H3Cl3O	1.47															19
1,2,4-Trichlorotoluene:	C7H5Cl3	1.375		2.08	-18												15,3
1,1,2-Trichloro trifluoroethane:	C2Cl3F3	1.564	-2.3	0.664	-12.6	0.911	2.1	76	-200	10.2	31	0.648	0.65	8.7		60	1,13,4
Tricresyl phosphate:	C12H21O4P	1.15		81.6	-4024	1.59											10,4
n-Tridecane:	C13H28	0.752	-0.7	1.746	-23.8	2.182	4.59	139	-160	8.8(300)	15	2.77(300)	2.9	30(300)		100	13,4
1-Tridecanol:	C13H28O	0.843		50(20)		2.365		160				1.45	4.29				10,4
1-Tridecene:	C13H26	0.764		1.45	-45	2.08	0	150	0			1.62					10,4
Triethanolamine:	C6H15NO3	1.1196	-0.54	613.6	-13759	2.06		196	66	5.6	23	1.599		8.5		55	10,14,4
Triethylamine:	C6H15N	0.721	-0.94	0.338	-4.8	2.236	5.5	120	-300	6.85	22	1.567	3.95	11.8		110	13,10,14,4
1,3,5-Triethylbenzene:	C12H18	0.854		1.698	-62.5	2	0	150	0								10
Triethylene glycol:	C6H14O4	1.12	-0.78	44	-1012	2.06											14,10,4
butyl ether:(mono)	C10H22O4	1.002		10.65													11
diacetate:	C10H18O6	1.119		11.35													11
diethyl ether:	C10H22O4																22
dimethyl ether:	C8H18O4	0.9862		3.55													11
ethyl ether:(mono)	C8H18O4	1.0208		4.25													11
methyl ether:(mono)	C7H16O4	1.049		7.02													11
Triethylenetetramine:	C6H18N4	0.982															15
Triethyl phosphate:	C6H15O4P																4

continued

Table Db-1 Selected Physical and Termal Properties of Gases and Liquids for Sizing Equipment (at 25°C unless noted)

	Formula	LIQUID								GAS						
		dens Mg/m	dD/dT x1000	vis mPa.s	dv/dT x1000	cp kJ/kg.K	dc/dT x1000	k mW/m.K	dk/dT x1000	vis μPa.s	dv/dT x1000	cp kJ/kg.K	dc/dT x1000	k mW/mK	dk/dt x1000	References
1,1,1-Trifluoroethane:	C2H3F3	0.99	-2.48	0.43(-50)		1.447	1.76	77.5	-540							13,4
Trifluoromethane: (R-14)	CHF3									14.8	48	0.73	1.35	15	70	13,4
Triglycol dichloride:	C6H14Cl2O2	1.2														23
1,3,5-Triisopropyl benzene:	C15H24	0.845														15
1,3,3-Trimethoxybutane:	C7H16O3	0.94														22,15
Trimethylamine:	C3H9N	0.631	-1.08	0.101	-3.76	2.315	3.68	117	-360	7.6	27	1.552	4.36	16	107	13,4
1,2,3-Trimethylbenzene:	C9H12	0.89	-0.78	0.956	-10.8	1.8	2.54	132	-200	(6.4)	19	(1.365)	2.89	(7.25)	90	13,4
2,2,3-Trimethylbutane:	C7H16	0.6901														14,4
1,1,2-Trimethylcyclopentane:	C8H16	0.7661														19
1,1,3-Trimethylcyclopentane:	C8H16	0.7703														19
2,2,3-Trimethylheptane:	C10H22															22
2,2,4-Trimethylheptane:	C10H22	0.7275														19
2,2,5-Trimethylheptane:	C10H22															4
2,2,6-Trimethylheptane:	C10H22															22
2,3,3-Trimethylheptane:	C10H22															22
2,3,4-Trimethylheptane:	C10H22															22
2,3,5-Trimethylheptane:	C10H22															22
2,3,6-Trimethylheptane:	C10H22															22
2,4,4-Trimethylheptane:	C10H22															22
2,4,5-Trimethylheptane:	C10H22															22
2,4,6-Trimethylheptane:	C10H22															22
2,5,5-Trimethylheptane:	C10H22															22
3,3,4-Trimethylheptane:	C10H22															22
3,3,5-Trimethylheptane:	C10H22	0.7248														19
3,4,4-Trimethylheptane:	C10H22															22
3,4,5-Trimethylheptane:	C10H22															22
2,2,3-Trimethylhexane:	C9H20															22
2,2,4-Trimethylhexane:	C9H20															22
2,2,5-Trimethylhexane:	C9H20	0.70322														14
2,3,3-Trimethylhexane:	C9H20															22
2,3,4-Trimethylhexane:	C9H20															22
2,3,5-Trimethylhexane:	C9H20															22
2,4,4-Trimethylhexane:	C9H20															22
3,3,4-Trimethylhexane:	C9H20															22
2,6,8-Trimethyl- -4-nonanol:	C12H26O	0.8193														23,14
2,2,3-Trimethylpentane:	C8H18	0.71207		0.566	-6.35							1.654				14,4
2,2,4-Trimethylpentane:	C8H18	0.687	-0.87	0.478	-3.4	2.088	4.4	96	-730			1.657				1,14,4
2,3,3-Trimethylpentane:	C8H18															4
2,3,4-Trimethylpentane:	C8H18															4
2,2,4-Trimethylpentanol:	C8H18O	0.8507														19
Trimethyl phosphate:	C3H9O4P															4
2,4,6-Trinitrotoluene:	C7H15N3O6	1.654														4,15
Triphenylmethane:	C19H16	1.014(400	-0.39	3.22(400)	-16.7	1.85(400)	0.62	136(400)	-40							13
Tripropylene glycol:	C9H20O4	1.442														10,15
-butyl ether:(mono)	C13H28O4	0.934														15
-ethyl ether:(mono)	C11H24O4	0.948														15
-isopropyl ether:(mono)	C12H26O4:	0.942														15
-methyl ether:(mono)	C10H22O4	0.967														15,23
Tryptophan:	C11H12N2O2															19
U*																
Uracil:	C4H4N2O2															19
Undecane:	C11H24	0.751	-0.67	1.49	-17.6	2.173	2.8	143	190	5.3	15	1.83	3.43	4.5	100	13,4
Undecanoic acid:	C11H22O2	0.9905		7.3(50)		2.26										5,25,21

Compound	Formula	Col 1	Col 2	Col 3	Col 4	Col 5	Col 6	Col 7	Col 8	Col 9	Col 10	Col 11	Col 12	Col 13	Col 14	Col 15	Ref.
1-Undecene:	C11H22	0.75															15,4
n-Undecyl alcohol:	C11H24O	0.8324															4
2-Undecyl alcohol:	C11H24O																
Uranium oxide:	U3O8	7.31															3
Urea:	CH4N2O	1.323															19,15,4
Urea nitrate:	CH5N3O4	1.69															19
Urethane:	C3H7NO2	1.056		5.23	-17												15
V*																	
Valeraldehyde:	C5H10O																4
Valine:	C5H11NO2	1.23															15
Veratrole:	C8H10O2	1.0819	-1	3.281	-52												14
Vinyl acetate:	C4H6O2	0.9247	-1.3	0.4	-3.8	1.67	1	143	-300	7.4	27	1	2.19	13		100	2,14,4
Vinyl allyl ether:	C8H16O2	0.8															23
Vinyl butyl ether:	C6H12O	0.7727		0.5(20)		2.32											14,4
Vinyl butyrate:	C6H10O2	0.9															23
Vinyl chloride:	C3H3Cl	0.8975	-1.78	0.43(-50)		1.361	0.04	114	-320	11	33	0.858	1.84	11		70	13,4
Vinyl 2-chloroethyl ether:	C4H7ClO																23
Vinyl crotonate:	C2H3Cl	0.9															23
Vinyl ethyl ether:	C4H8O	0.747	-1.17	0.2(20)													14,4
Vinyl 2-ethylhexanoate:	C10H18O2	0.9															23
Vinyl 2-ethylhexyl ether:	C10H20O																23
Vinyl isobutyl ether:	C6H12O	0.8															23
Vinyl isopropyl ether:	C5H10O																23
Vinyl methyl ether:	C3H6O																4
Vinyl propionate:	C5H8O2																4
m-Vinyltoluene:	C9H10			0.791	n=-3.38												14
o-Vinyltoluene:	C9H10																14
p-Vinyltoluene:	C9H10																4
W*																	
Water:	H2O	0.99707	-0.42	0.89	-9.8	4.186	-0.8	608	1028	9.35	39	2.042	0	18.93		82	13,4,1,14
				0.279*	-0.16*	4.216*	2.8*	680*	-160*	12.1*	35*	2.038*	8.4*	24.8*	127*		
Water, heavy:	D2O	1.103	-0.53	1.195	-11.6	4.22	0	583	600								13
X*																	
Xenon:	Xe									22.9	72	0.159	0	6		20	13
m-Xylene:	C8H10	0.8617	-0.86	0.579	-7.14	1.733	3.12	131	-240	(6.5)	20	(1.31)	2.94	(5.5)		100	13,4,1
		0.752*		0.232*		2.13*		92*		9.1*		1.66*		16.9*			
o-Xylene:	C8H10	0.8759	-1.02	0.753	-10.9	1.749	3.12	133	-180	(6.9)	19	(1.358)	2.84	(6.5)		100	13,4,1
		0.764*		0.253*		2.16*		99*		9.2*		1.71*		18.4*			
p-Xylene:	C8H10	0.857	-0.88	0.621	-4.4	1.716	3.23	131	-240	(6.59)	20	(1.300)	2.93	(5.5)		100	13,4,1
		0.753*		0.232*		2.11*		98*		8.8*		1.64*		17.1*			

For liquid viscosity, the temperature variation is probably best represented by an Othmer type or Cox Chart type representation with a log viscosity plotted against a temperature scale created from the log vapor pressure of water. In keeping with the principle of optimum sloppiness, this degree of accuracy is inconsistent with the time we have to estimate the property. Hence, the two simple methods of reporting property variation with temperature are: (for liquid viscosity **only**)

a linear variation $\mu_T = \mu_{25} + (T - 298)\, d\mu/dT \times 10^{-3}$ (Db-1)
an exponential variation $\mu_T = \mu_{25} \times (T/298)^n$ (Db-2)

For the exponential variation, the table entry is given as n = . . . and the usual value is –4.

The greatest challenge in compiling these data was terminology. In general, I tried to use the terminology used by Danbert and Danner (1985) and updates since this is the follow-up resource we would probably use for more accurate estimates. To facilitate this use, the chemical formula is also given. To guide in the use of terminology, synonyms, trade names, common names, and Geneva convention names of compounds, use the Chemical Index first. For example, "Cellosolve" is "the ethyl ether of ethylene glycol," "Jefferson EE," or "2-ethoxyethanol." In this table, this is located under "Ethylene glycol, ethyl ether (mono)." To find your way through the maze of terminology, consult the Chemical Index first.

Surprisingly few data are available for the properties of species. The available data vary. Where there was variability, the preference was for experimentally reported data before estimations; more recent and from sources that seemed to be consistently reliable. On this basis, the data were selected from Vargaftik (1975), Beaton and Hewitt (1989), Dean (1987), Riddick and Bunger (1970), Ullmann (1985), and Danbert and Danner (1985). In addition, trade literature (designated T) provided a lot of otherwise unavailable data.

Details of methods of estimation the properties of pure components and of mixtures are given by Reid, Prausnitz and Sherwood (1977). Rapid estimation of the thermal conductivity for liquids can be done from Figure Db–2.

REFERENCES FOR PART Da AND b:

1. VARGAFTIK, N. B. 1975. *Handbook of Physical Properties of Liquids and Gases,* 2nd ed. Washington DC: Hemisphere Publishing Co. Good starting reference. All the data are in SI already.
2. GALLANT, R. W., and J. M. RAILEY. 1984. *Physical Properties of Hydrocarbons,* Vol. 2, 2d ed. Houston, TX: Gulf Publishing Co.
3. PERRY'S *Chemical Engineer's Handbook.* New York: McGraw-Hill.
4. DANBERT, T. E., and R. P. DANNER. 1985. *Data Compilation Tables of Properties of Pure Compounds.* New York: DIPPR and AIChE.
5. MCKETTA'S *Encyclopedia of Chemical Process Design.* New York: Marcel Dekker. (list by component name)
6. LANGE'S *Handbook of Chemistry.* 13th ed. New York: McGraw-Hill.
7. YAWS, CARL L., et al. 1977. "Physical Properties." *Chem Eng.* This book is a republishing of a series of paper that appeared in *Chem Eng* magazine. The data are reasonably complete although they include only about 80 compounds.
8. DREISBACH, R. R. 1961. "Physical Properties of Chemical Compounds III," ACS No. 29, Am. Chem Soc., Washington DC, provides tables for each compound;
9. AERSTIN, F., and G. STREET. 1978. *Applied Chemical Process Design.* New York: Plenum Press.
10. DE RENZO, D. J. *Solvents Safety Handbook.* Park Ridge, NJ: Noyes Data.
11. *Kirk-Othmer Encyclopedia of Chemical Technology.* M. Grayson, ed., 3rd ed., A Wiley-Interscience Publication. New York: John Wiley & Sons.
12. MARSDEN, C., and S. MANN 1963. *Marsden Solvent Guide.* London: Cleaver-Hume Press.
13. BEATON, C. F., and G. F. HEWITT. 1989. *Physical Property Data for the Design Engineer.* New York: Hemisphere Publishing.
14. RIDDICK, J. A., and W. B. BUNGER. 1970. *Techniques of Chemistry, Vol II, Organic Solvents.* 3rd ed. New York: Wiley Interscience,
15. DEAN, J. A. 1987. *Handbook of Organic Chemistry.* New York: McGraw-Hill.
16. STEERE, N. V. 1971. *CRC Handbook of Laboratory Safety,* 2nd ed. West Palm Beach, FL: CRC Press.
17. DUHNE, C. R. 1979. "Viscosity Temperature Correlations for Liquids." *Chemical Engineering* **86,** 15 (July 16):83–91.
18. YAWS, C. L. series of articles in *Chemical Engineering* and in *Hydrocarbon Processing.* For example, "Calculate Liquid Heat Capacity" *Hydrocarbon Process* (Dec 1991):73–77; or "Heat Capacities for 700 Compounds" by Yaws, H. M. Ni, and P. Y. Chiang, *Chemical Engineering* (May 9, 1988):91–98.
19. LIDE, D. R. *Handbook of Chemistry and Physics* 1990–91, 71st edition. Boca Raton, FL: CRC Press.
20. ALDICH'S 1988. *Catalog Handbook of Fine Chemicals.* Milwaukee, WI: Aldich Chemical Co.
21. SWERN, D. 1964. *Bailey's Industrial Oil and Fat Products,* 3rd ed. New York: Wiley Interscience.
22. BARTON, A. F. M. *Solubility Parameters.* Cleveland, OH: CRC Press. (Prime source of the solubility parameters.)
23. MCKINNON, G. P, and K. TOWER. 1976. *Fire Protection Handbook.* 14th ed. Boston, MA: National Fire Protection Association.
24. BUCKINGHAM, J. 1982. *Dictionary of Organic Compounds.* 5th ed. New York: Chapman Hall,
25. RALSTON, A. W. 1948. *Fatty Acids and Their Derivatives.* New York: J. Wiley and Sons.
26. ULLMANN 1985. *Ullmann's Encyclopedia of Industrial Chemistry.* 5th ed. New York: VCH, Weinheim.

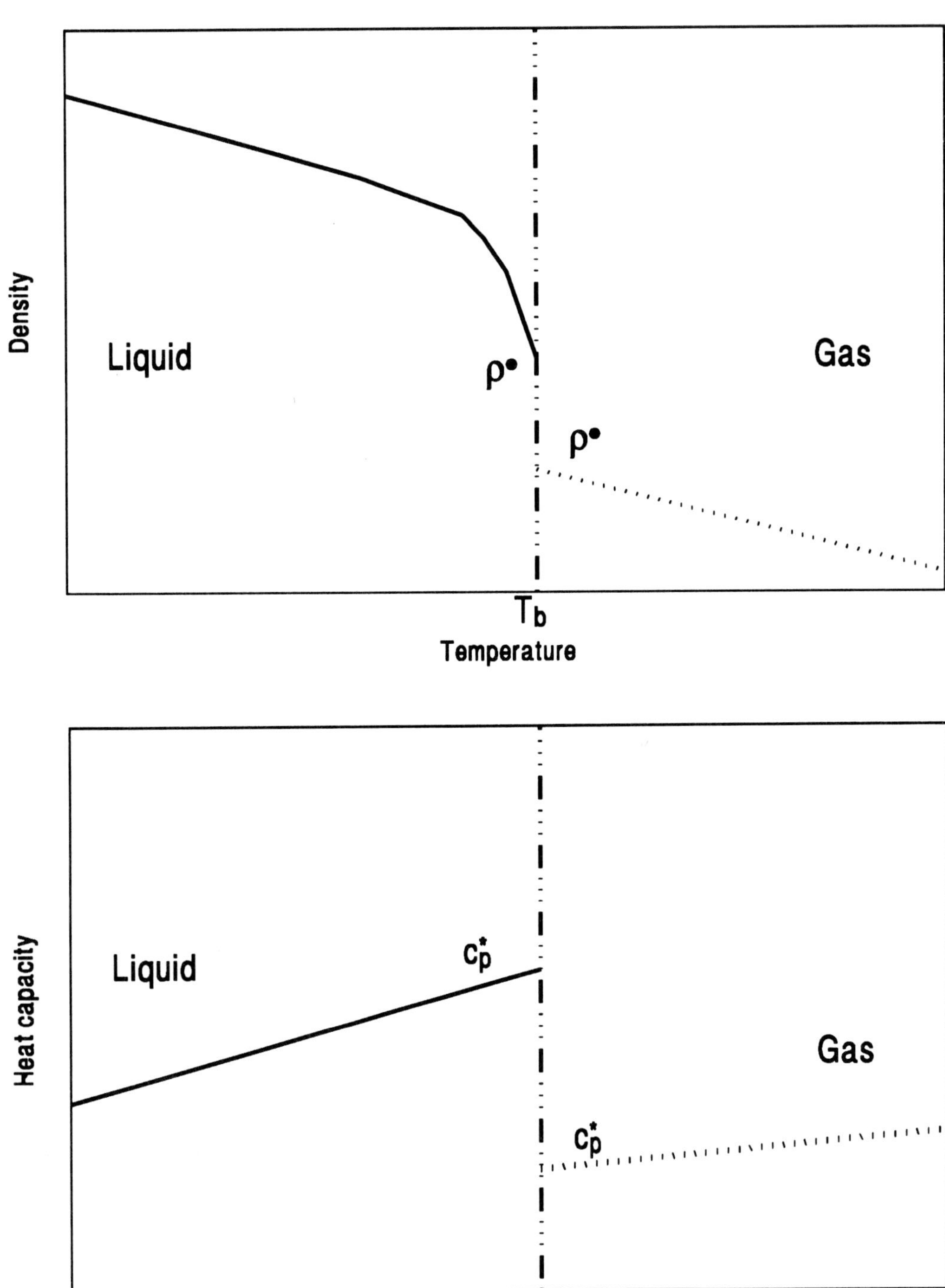

Figure Db–1 Property Variation with Temperature
a) Density
b) Heat Capacity

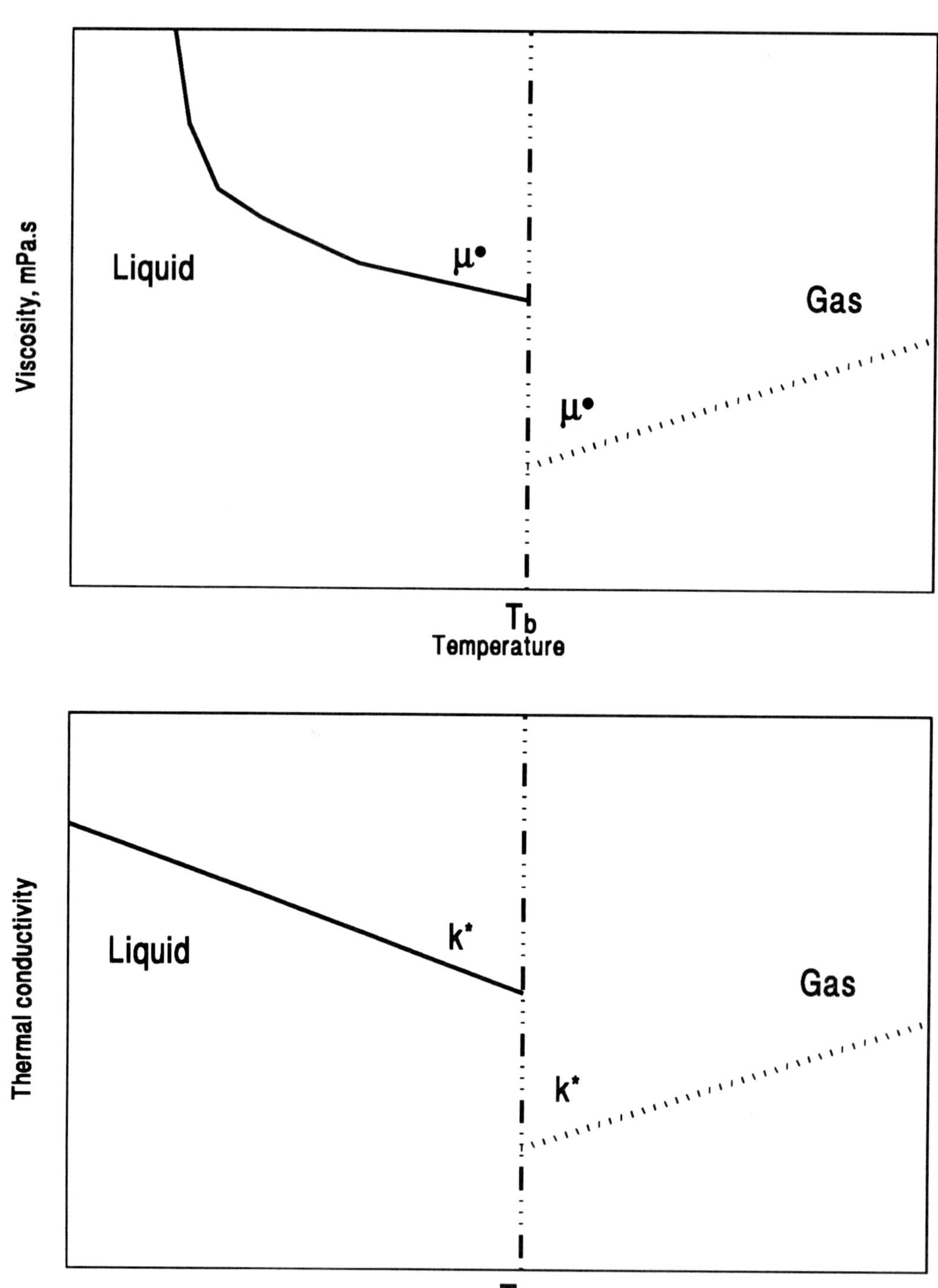

Figure Db–1 Property Variation with Temperature
c) Viscosity
d) Thermal Conductivity

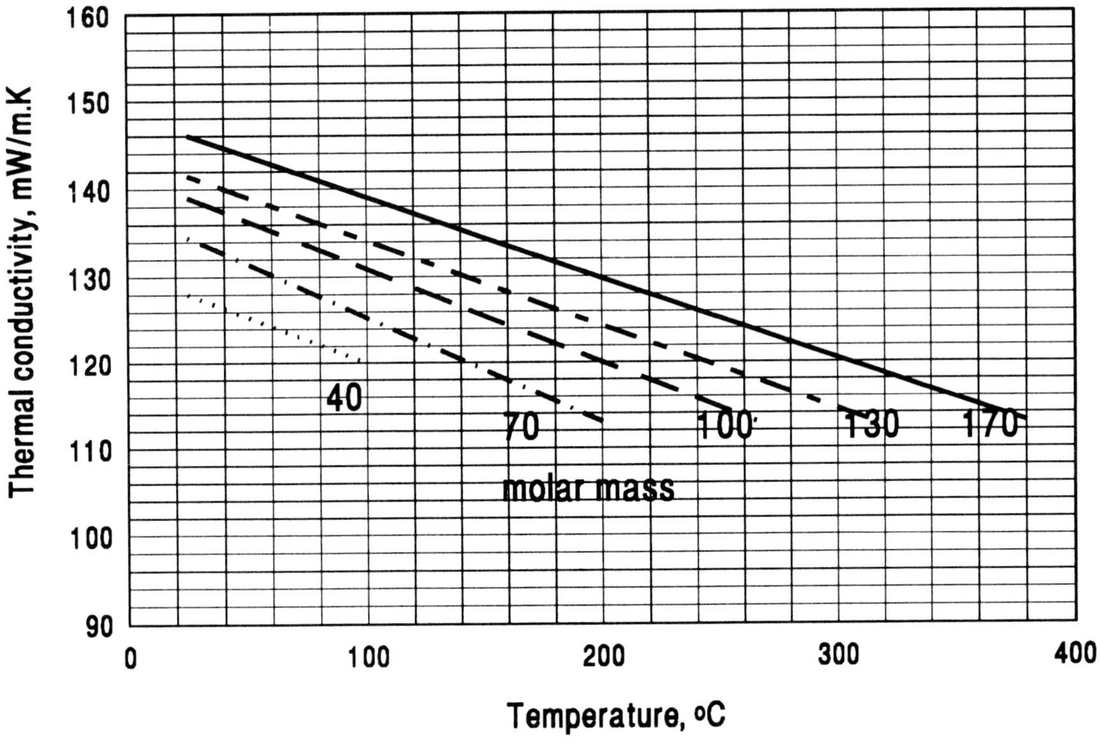

Figure Db–2 Estimating the Thermal Conductivity of Liquid Organics

27. Dow's Fire and Explosive Index Hazard Classification Guide. 5th ed. 1981. New York: Amer Institute of Chemical Engineers.

28. DE RENZO, D. J. 1980. *Biodegradation Techniques for Industrial Organic Wastes.* Park Ridge, NJ: Noyes Data Corporation. One of the few resources available for data about the biodegradability. Cites data of Pitter (1976).

29. PITTER, P. 1976. "Determination of Biological Degradability of Organic Substances." *Water Research* **10:**231–235. (Activated sludge conditions)

30. DESHPAND, S. D. 1987. "Activated sludge conditions: TOC degradation data for resorcinol, catechol, benzensulfonic acid, m-nitrobenzenesulfonate and m-aminophenol." *Environmental Sci and Tech* **21:**1003.

31. TABAK, H. H., et al. 1981. "Biodegradability studies with organic priority pollutant compounds." *Journal WPCF* **53,** 10:1503–1518.

32. MARTELL, A. E., and R. M. SMITH. 1979. *Critical Stability Constants.* New York: Plenum Press. dissociation constants for species in water. Data seem to be summarized in Lange's *Handbook.*

33. STUMM, W., and J. J. MORGAN. 1981. "Aquatic Chemistry." 2nd ed. New York: J. Wiley and Sons. Solubility constants: pp. 241 ff.

34. HAMER, P., et al. 1961. *Industrial Water Treatment Practice.* London: Butterworths. Solubility constants: pp. 420 ff.

35. MULLIN, J. W. 1972. *Crystallization.* 2nd ed. London: Butterworths. Solubility constants: p. 419. These are more for the hydrated species.

36. SKOOG, D. A., and D. M. WEST. 1976. *Fundamentals of Analytical Chemistry.* 4th ed. New York: Holt, Rinehart and Winston. Solubility products: p. 782; Dissociation constants pp. 784 ff.

37. REID, R. C., J. M. PRAUSNITZ, and T. K. SHERWOOD. 1977. *The Properties of Gases and Liquids.* 3rd ed. New York: McGraw-Hill; 4th edition 1987 with Reid, Prausnitz, and B. E. Poling.

Chemical Index

Some data are available in the companion book **Process Design and Engineering Practice.** Page numbers appearing in this index with the prefix "1-, 2-, 3-, 4-, 5-, 6- and A-" refer to data in **PDEP.**

This index identifies the compound name used and the location for properties for that compound. The code is:

Solids: size, hazard and characteristics, bulk density, conveyability parameters, angle of repose, and design data for hydraulic and pneumatic conveying. The numerical values may not be included for all compounds; however, the Table format is designed so that data for similar compounds can be located. Table C-3, pages 28 to 38. Listed alphabetically.

! Vapour pressure, Figure C-9, pages 54 and 55.

@ Solids: formula, mass density, hardness, work index, zpc, relative permittivity, threshold voltage for electrostatic charging and magnetic properties. Part C-4, pages 39 to 45. Listed alphabetically.

* Liquids and gases: major resources. Table Da-1 gives molar mass, NFPA rating, Dow hazard factor, corrosivity of copper, carbon steel and 316 stainless steel, freezing and boiling temperatures, molar volumes, Hidebrand polar and hydrogen solubility parameters, dissociation constant, solubility product, biodegradability and the latent heats of melting and vaporization, 58 to 82. Listed alphabetically. For the same compounds, Table Db-1 gives the property at 25°C and its temperature variation: liquid density, gas and liquid viscosities, gas and liquid heat capacities, and gas and liquid thermal conductivities. pages 83 to 108.

~ Production capacity of single plants, Table 4-4, page 4-18 in PDEP.

| Absorption characteristics of the gaseous species by a variety of solvents, Table 1B-2, page 1-25 ff, in PDEP.

^ Density of a liquid solution, Fig B-1, page 21.

Density of a gas can be estimated from the ideal gas law, or from Figure C-6, page 49.

B

solids, 28 ff; @ solids, 39 ff; ! vapour pressure Fig C-9; * separation criteria, 58 to 82; transport properties; 83 to 108; ! vapour pressure; ~ production capacity, 4-18, in PDEP; | absorption, 1-25, in PDEP; ^ solution density, 21.

solids, 28 ff; @ solids, 39 ff; ! vapour pressure Fig C-9; * separation criteria, 58 to 82; transport properties; 83 to 108; ! vapour pressure; ~ production capacity, 4-18, in PDEP; | absorption, 1-25, in PDEP; ^ solution density, 21.

D*

solids, 28 ff; @ solids, 39 ff; ! vapour pressure Fig C-9; * separation criteria, 58 to 82; transport properties; 83 to 108; ! vapour pressure; ~ production capacity, 4-18, in PDEP; | absorption, 1-25, in PDEP; ^ solution density, 21.

E*

solids, 28 ff; @ solids, 39 ff; ! vapour pressure Fig C-9; * separation criteria, 58 to 82; transport properties; 83 to 108; ! vapour pressure; ~ production capacity, 4-18, in PDEP; | absorption, 1-25, in PDEP; ^ solution density, 21.

H*

solids, 28 ff; @ solids, 39 ff; ! vapour pressure Fig C-9; * separation criteria, 58 to 82; transport properties; 83 to 108; ! vapour pressure; ~ production capacity, 4-18, in PDEP; | absorption, 1-25, in PDEP; ^ solution density, 21.

I*

K*

L*

M*

solids, 28 ff; @ solids, 39 ff; ! vapour pressure Fig C-9; * separation criteria, 58 to 82; transport properties; 83 to 108; ! vapour pressure; ~ production capacity, 4-18, in PDEP; | absorption, 1-25, in PDEP; ^ solution density, 21.

N*

solids, 28 ff; @ solids, 39 ff; ! vapour pressure Fig C-9; * separation criteria, 58 to 82; transport properties; 83 to 108; ! vapour pressure; ~ production capacity, 4-18, in PDEP; | absorption, 1-25, in PDEP; ^ solution density, 21.

O*

P

solids, 28 ff; @ solids, 39 ff; ! vapour pressure Fig C-9; * separation criteria, 58 to 82; transport properties; 83 to 108; ! vapour pressure; ~ production capacity, 4-18, in PDEP; | absorption, 1-25, in PDEP; ^ solution density, 21.

Q*

R*

S*

solids, 28 ff; @ solids, 39 ff; ! vapour pressure Fig C-9; * separation criteria, 58 to 82; transport properties; 83 to 108; ! vapour pressure; ~ production capacity, 4-18, in PDEP; | absorption, 1-25, in PDEP; ^ solution density, 21.

T*

solids, 28 ff; @ solids, 39 ff; ! vapour pressure Fig C-9; * separation criteria, 58 to 82; transport properties; 83 to 108; ! vapour pressure; ~ production capacity, 4-18, in PDEP; | absorption, 1-25, in PDEP; ^ solution density, 21.

W*

X*

Y

Z

Annotated Subject Index

H

I

K

L

M

N

O

P

R

S